ESP 实用视听说系列教材

【总主编 赵团结】

食品化工英语

主 编 黄 迎

副主编 肖 灿 赵团结

审 校 聂勇伟 赵团结

Food &
Chemistry
English

首都经济贸易大学出版社

Capital University of Economics and Business Press

·北 京·

图书在版编目(CIP)数据

食品化工英语/黄迎主编. --北京:首都经济贸易大学出版社,2020.2

ISBN 978-7-5638-3061-9

I.①食… II.①黄… III.①食品化学—化学工程—英语—教材 IV.①TS201.2

中国版本图书馆 CIP 数据核字(2020)第 010638 号

食品化工英语

主　编　黄　迎

副主编　肖　灿　赵团结

Shipin Huagong Yingyu

责任编辑　胡　兰

封面设计　**风得信·阿东**　FondesyDesign

出版发行　首都经济贸易大学出版社

地　　址　北京市朝阳区红庙 (邮编 100026)

电　　话　(010)65976483　65065761　65071505(传真)

网　　址　http://www.sjmcb.com

E-mail　publish@cueb.edu.cn

经　　销　全国新华书店

照　　排　北京砚祥志远激光照排技术有限公司

印　　刷　北京玺诚印务有限公司

开　　本　787 毫米×1092 毫米　1/16

字　　数　377 千字

印　　张　14.75

版　　次　2020 年 2 月第 1 版　2020 年 2 月第 1 次印刷

书　　号　ISBN 978-7-5638-3061-9

定　　价　39.00 元

序 言
PREFACE

改革开放以后，为加强对外交往，英语被国家教育部门确定为大学必修课程。从此，万千中国学子投入大量的时间与精力去学习英语。这种长时间的重视与投入，极大地提高了国民的英语水平，密切了中外交往，加快了中国现代化进程，从而有力地促进了中国特色社会主义伟大事业的发展。经历了逾40年改革开放风雨洗礼的中国，正逐步走上世界舞台的中央。越来越有中国特色社会主义道路自信、理论自信、制度自信、文化自信的中国人民开始重新审视这种不分专业、齐头并进的全民大学英语教育，在客观、理性的探讨中已逐步形成共识，那就是：大学英语教育必须改革。可是，该如何改革呢？

回望过去是为了规划未来！中华儿女学习英语的热情在学习西方先进科技、开放图强的怒吼中被点燃，在追赶西方、复兴文化的脚步声中日渐高涨。时至今日，我们的英语学习怎能忘却初心、迷失方向，甚或不自觉地成为自毁文化长城的西方特洛伊木马？我们的大学英语教育改革必须把正航向。

知易行难！"雄关漫道真如铁，而今迈步从头越！"武昌工学院是以培养具有创新精神的应用型人才为己任的工科院校，必须以舍我其谁的气魄，在大学英语教学改革中勇当开路先锋。为此，对武昌工学院的大学英语教学改革，我提出了落实"两个一千"（1 000个专业词汇，1 000个专业句子）的具体目标。"功崇惟志，业广惟勤"，武昌工学院全体英语教师，在赵团结同志的带领下，志存高远，攻坚克难，积极推进 ESP（English for Specific Purposes）大学英语教学改革，花费大量心血，在深入调查研究的基础上，精心编写了这套"ESP实用视听说系列教材"。"德不孤，必有邻"，我相信，这套涵盖七大专业门类的英语教材必能引起同行的关注，得到他们的爱护、响应与支持。"大厦之成，非一木之材也；大海之阔，非一流之归也"，希望本套英语教材的全体编写同志，戒骄戒躁，汇智聚力，与兄弟院校的同行携手并进，不断提高教材质量，悉心将它打造成业界精品，共同回应时代要求，开辟大学英语教学改革的新天地！

无尽今来古往，多少春花秋月！任何事业都是在后人赶超前人的奋斗中，不断自我更新扬弃，从而长盛不衰的。"芳林新叶催陈叶，流水前波让后波"，英语教学改革永远在路上。希望同志们不断赶超前人，持续推进英语教学改革。"合抱之木，生于毫末；九层之台，起于累土。"ESP大学英语教学改革，已经迈出了坚实的第一步，必能善始善终，行稳致远，收前人未竟之功。

是为序。

武昌工学院董事长兼校长　李勇

2019 年 12 月 15 日

本书使用说明
INSTRUCTIONS

　　"ESP 实用视听说系列教材"是武昌工学院国际教育学院组织编写的一套教材,适用于普通高等院校本科专业大三的学生。该系列教材以学生专业学习阶段的专业词汇和专业句子为基础,以视听和口语练习为主要教学内容,强化学生的专业英语应用能力。本系列教材共 7 个分册,分别是:《机械工程英语》《土木工程英语》《信息工程英语》《食品化工英语》《艺术设计英语》《经济管理英语》《财务会计英语》。每个分册包括 10 个单元,覆盖相关专业领域内的主要专业或者方向。选用这套教材的学校,可以根据学生的专业灵活调整单元的顺序,自由选择教学的内容。

　　每个单元包括六个模块,分别是:①Pre-class Activity;②Specialized Terms;③Watching and Listening;④Talking;⑤After-class Exercises;⑥Additional Reading。第一个模块是课前活动,为学生简要介绍相关领域里的名人,让学生预习,并进行自由讨论,作为课前热身。第二个模块要求学生课前预习专业词汇,教师可以在课堂上采用听写、词语接龙等形式检查学生对专业词汇的记忆效果。第三个模块是视听训练,所选视频均来自国外主流网站最新的视频资料,并根据学生的接受程度,根据由易到难的原则,编写了选择题、判断题、填空题和讨论题。第四个模块是口语训练,包括需要学生诵读的 100 个经典句子(50 个通用句子和 50 个专业句子),以及供班级组织活动的两个对话样板,然后进行模仿练习和较高难度的讨论和辩论练习。第五个模块主要是结合本单元视听材料中出现的专业词汇,编写了配对、填空、翻译、写作等相关应用题型。第六个模块是课后阅读,主要内容是相关领域内的世界著名公司的简介,并采用四级考试中的长篇阅读题型,加强学生信息捕获的能力。

　　每个单元大约需要 6 个学时。建议第一、二模块用 0.5 个学时,第三模块大概需要 2.5 个学时,第四模块需要 2 个学时,第五、六模块需要 1 个学时。为了方便师生使用本教材,我们把视频文字和练习答案,放在了出版社的网络平台上,扫描书中的二维码,即可查阅;我们还提供了相关听力视频的网址供广大师生查阅。如果无法在网上查到视频或有任何疑问和批评建议,可以联系本套丛书的总主编赵团结教授,他的联系方式如下:

　　手机:13986141410
　　座机:027-88142008
　　邮箱:markztj@ sina.com
　　QQ 号:3153144383
　　微信号:13986141410

编者

2020 年 1 月 3 日

CONTENTS

Unit One Chemical Elements

I. Pre-class Activity

Directions: *Please read the general introduction about Madame Curie and tell something more about the great scientist to your classmates.*

Madame Curie

Mary Curie（1867-1934）is also known as "Madame Curie", with a full name of Maria Schwadovska Curie. As a famous French Poland scientist, physicist and chemist, she was born on November 7, 1867 in Warsaw.

In 1903, Marie Curie and Becquerel, due to the study of radioactivity（放射性）, were jointly awarded the Nobel Prize of physics. In 1911, for the discovery of polonium（钋）and radium（镭）, she again won the Nobel Prize in Chemistry, making her become the first scientist who won the Nobel Prize twice. Madame Curie's achievements include having created a radioactive theory, having invented the technology of separating radioactive isotope（放射性同位素）, and having discovered two new elements, namely polonium and radium. Under her guidance, it was the first time for people to apply the radioactive isotope to the treatment of cancer. Due to long-term exposure to radioactive substances, Mrs. Curie died of malignant leukemia（恶性白血病）on July 3, 1934.

II. Specialized Terms

Directions: *Please memorize the following specialized terms before the class so that you will be able to better cope with the coming tasks.*

actinium n. 锕（第 89 号元素）：[化学符号]Ac, 读"阿"

aluminum n. 铝（第 13 号元素）：[化学符号]Al, 读"吕"

americium n. 镅(第 95 号元素):[化学符号]Am,读"眉"

antimony n. 锑(第 51 号元素):[化学符号]Sb,读"梯"

argon n. 氩(第 18 号元素):[化学符号]Ar,读"亚"

arsenic n. 砷(第 33 号元素):[化学符号]As,读"申"

astatine n. 砹(第 85 号元素):[化学符号]At,读"艾"

barium n. 钡(第 56 号元素):[化学符号]Ba,读"贝"

berkelium n. 锫(第 97 号元素):[化学符号]Bk,读"陪"

beryllium n. 铍(第 04 号元素):[化学符号]Be,读"皮"

bismuth n. 铋(第 83 号元素):[化学符号]Bi,读"必"

bohrium n. 铍(第 107 号元素):[化学符号]Bh,读"波"

boron n. 硼(第 05 号元素):[化学符号]B,读"朋"

bromine n. 溴(第 35 号元素):[化学符号]Br,读"秀"

cadmium n. 镉(第 48 号元素):[化学符号]Cd,读"隔"

calcium n. 钙(第 20 号元素):[化学符号]Ca,读"丐"

californium n. 锎(第 98 号元素):[化学符号]Cf,读"开"

carbon n. 碳(第 06 号元素):[化学符号]C,读"炭"

cerium n. 铈(第 58 号元素):[化学符号]Ce,读"市"

cesium n. 铯(第 55 号元素):[化学符号]Cs,读"色"

chlorine n. 氯(第 17 号元素):[化学符号]Cl,读"绿"

chromium n. 铬(第 24 号元素):[化学符号]Cr,读"各"

cobalt n. 钴(第 27 号元素):[化学符号]Co,读"古"

copper n. 铜(第 29 号元素):[化学符号]Cu,读"同"

curium n. 锔(第 96 号元素):[化学符号]Cm,读"局"

darmstadtium n. 鐽(第 110 号元素):[化学符号]Ds,读"达"

dubnium n. 铳(第 105 号元素):[化学符号]Db,读"杜"

dysprosium n. 镝(第 66 号元素):[化学符号]Dy,读"滴"

einsteinium n. 锿(第 99 号元素):[化学符号]Es,读"哀"

erbium n. 铒(第 68 号元素):[化学符号]Er,读"耳"

europium n. 铕(第 63 号元素):[化学符号]Eu,读"有"

fermium n. 镄(第 100 号元素):[化学符号]Fm,读"费"

fluorine n. 氟(第 09 号元素):[化学符号]F,读"弗"

francium n. 钫(第 87 号元素):[化学符号]Fr,读"方"

gadolinium n. 钆(第 64 号元素):[化学符号]Gd,读"嘎(二声)"

gallium n. 镓(第 31 号元素):[化学符号]Ga,读"家"

germanium n. 锗(第 32 号元素):[化学符号]Ge,读"者"

gold n. 金(第 79 号元素):[化学符号]Au,读"今"

hafnium n. 铪(第 72 号元素):[化学符号]Hf,读"哈"

hassium n. 镙(第 108 号元素):[化学符号]Hs,读"黑"

helium n. 氦(第 02 号元素)：[化学符号]
He,读"亥"

holmium n. 钬(第 67 号元素)：[化学符
号]Ho,读"火"

hydrogen n. 氢(第 01 号元素)：[化学符
号]H,读"轻"

indium n. 铟(第 49 号元素)：[化学符号]
In,读"因"

iodine n. 碘(第 53 号元素)：[化学符号]
I,读"典"

iridium n. 铱(第 77 号元素)：[化学符号]
Ir,读"衣"

iron n. 铁(第 26 号元素)：[化学符号]
Fe,读"铁"

krypton n. 氪(第 36 号元素)：[化学符
号]Kr,读"克"

lanthanum n. 镧(第 57 号元素)：[化学符
号]La,读"蓝"

lawrencium n.铹(第 103 号元素)：[化学
符号]Lw,读"劳"

lead n. 铅(第 82 号元素)：[化学符号]
Pb,读"千"

lithium n. 锂(第 03 号元素)：[化学符号]
Li,读"里"

lutetium n. 镥(第 71 号元素)：[化学符
号]Lu,读"鲁"

magnesium n. 镁(第 12 号元素)：[化学符
号]Mg,读"美"

manganese n. 锰(第 25 号元素)：[化学符
号]Mn,读"猛"

meitnerium n. 䥑(第 109 号元素)：[化学
符号]Mt,读"麦"

mendelevium n. 钔(第 101 号元素)：[化
学符号]Md,读"门"

mercury n. 汞(第 80 号元素)：[化学符
号]Hg,读"拱"

molybdenum n. 钼(第 42 号元素)：[化学
符号]Mo,读"目"

neodymium n. 钕(第 60 号元素)：[化学符
号]Nd,读"女"

neon n. 氖(第 10 号元素)：[化学符号]
Ne,读"乃"

neptunium n. 镎(第 93 号元素)：[化学符
号]Np,读"拿"

nickel n. 镍(第 28 号元素)：[化学符号]
Ni,读"臬"

niobium n. 铌(第 41 号元素)：[化学符
号]Nb,读"尼"

nitrogen n. 氮(第 07 号元素)：[化学符
号]N,读"淡"

nobelium n. 锘(第 102 号元素)：[化学符
号]No,读"诺"

osmium n. 锇(第 76 号元素)：[化学符
号]Os,读"鹅"

oxygen n. 氧(第 08 号元素)：[化学符号]
O,读"养"

palladium n. 钯(第 46 号元素)：[化学符
号]Pd,读"巴"

phosphorus n. 磷(第 15 号元素)：[化学符
号]P,读"邻"

platinum n. 铂(第 78 号元素)：[化学符
号]Pt,读"博"

plutonium n. 钚(第 94 号元素)：[化学符
号]Pu,读"布"

polonium n. 钋(第 84 号元素)：[化学符
号]Po,读"泼"

potassium n. 钾(第 19 号元素)：[化学符
号]K,读"甲"

praseodymium n. 镨(第 59 号元素)：[化
学符号]Pr,读"普"

promethium n. 钷(第 61 号元素)：[化学
符号]Pm,读"颇"

protactinium n. 镤(第 91 号元素)：[化学
符号]Pa,读"葡"

radium n. 镭(第 88 号元素)：[化学符号]
Ra,读"雷"

radon n. 氡(第 86 号元素):[化学符号]Rn,读"冬"

rhenium n. 铼(第 75 号元素):[化学符号]Re,读"来"

rhodium n. 铑(第 45 号元素):[化学符号]Rh,读"老"

roentgenium n. 轮(第 111 号元素):[化学符号]Rg,读"伦"

rubidium n. 铷(第 37 号元素):[化学符号]Rb,读"如"

ruthenium n. 钌(第 44 号元素):[化学符号]Ru,读"了"

rutherfordium n. 铲(第 104 号元素):[化学符号]Rf,读"卢"

samarium n. 钐(第 62 号元素):[化学符号]Sm,读"衫"

scandium n. 钪(第 21 号元素):[化学符号]Sc,读"亢"

seaborgium n. 镭(第 106 号元素):[化学符号]Sg ,读"喜"

selenium n. 硒(第 34 号元素):[化学符号]Se,读"西"

silicon n. 硅(第 14 号元素):[化学符号]Si,读"归"

silver n. 银(第 47 号元素):[化学符号]Ag,读"银"

sodium n. 钠(第 11 号元素):[化学符号]Na,读"纳"

strontium n. 锶(第 38 号元素):[化学符号]Sr,读"思"

sulfur n. 硫(第 16 号元素):[化学符号]S,读"流"

tantalum n. 钽(第 73 号元素):[化学符号]Ta,读"坦"

technetium n. 锝(第 43 号元素):[化学符号]Tc,读"得"

tellurium n. 碲(第 52 号元素):[化学符号]Te,读"帝"

terbium n. 铽(第 65 号元素):[化学符号]Tb,读"特"

thallium n. 铊(第 81 号元素):[化学符号]Tl,读"他"

thorium n. 钍(第 90 号元素):[化学符号]Th,读"土"

thulium n. 铥(第 69 号元素):[化学符号]Tm,读"丢"

tin n. 锡(第 50 号元素):[化学符号]Sn,读"西"

titanium n. 钛(第 22 号元素):[化学符号]Ti,读"太"

tungsten n. 钨(第 74 号元素):[化学符号]W,读"乌"

uranium n. 铀(第 92 号元素):[化学符号]U,读"由"

vanadium n. 钒(第 23 号元素):[化学符号]V,读"凡"

xenon n. 氙(第 54 号元素):[化学符号]Xe,读"仙"

ytterbium n. 镱(第 70 号元素):[化学符号]Yb,读"意"

yttrium n. 钇(第 39 号元素):[化学符号]Y,读"乙"

zinc n.锌(第 30 号元素):[化学符号]Zn,读"辛"

zirconium n. 锆(第 40 号元素):[化学符号]Zr,读"告"

III. Watching and Listening

Task One　Elements and Atoms（Ⅰ）

New Words

atom n.原子

property n.特性,属性

solid adj.固体的

liquid n.液体 adj. 液体的

graphic adj.图解的

particle n.微粒

gassy adj. 气体的;气状的

philosophical adj.哲学的

chunk n.块

substance n.物质,材料

terminology n.术语

periodic adj.周期的

unbelievably adv.难以置信地

cross-section n.横断面;横截面图

string n.绳子 v. 绑,系或用线挂起

width n.宽度

pluck v.拔掉

block n.块

视频链接及文本

Exercises

1. *Watch the video for the first time and choose the best answers to the following questions.*

1）Which of the following is not an element in the field of chemistry?

　A. Lead.　　　　　　　　　　B. Water.

　C. Carbon.　　　　　　　　　D. Gold.

2）Every substance can exist in all the following states except _____.

　A. solid　　　　　　　　　　B. liquid

　C. gas　　　　　　　　　　　D. earth

3）Which is the smallest unit of substance in nature?

　A. Element.　　　　　　　　　B. Atom.

　C. Molecule.　　　　　　　　　D. Dust.

4）When _____ is proper enough,every substance in nature can turn into liquid or gas.

　A. temperature　　　　　　　　B. humidity

　C. air　　　　　　　　　　　　D. water

5）Which of the following is not a metal?

　A. Lead.　　　　　　　　　　B. Gold.

　C. Silicon.　　　　　　　　　D. Iron.

2. *Watch the video again and decide whether the following statements are true or false.*

1）Nowadays,water is still considered as an element in the field of chemistry.（　）

2）Water is composed of nitrogen and oxygen.（　）

3）In the periodic table of elements,Au stands for gold.（　）

4）Nitrogen,oxygen and hydrogen are all gases under normal temperature.（　）

5) At the cross-section of a hair, there are only one thousand carbon atoms.(　　)

3. *Watch the video for the third time and fill in the following blanks.*

Just looking at the environment around us, there are different 1) _____ and these substances tend to have different 2) _____ . One may reflect the 3) _____ in different ways. One may be a certain 4) _____ at a certain temperature. We also start to observe how they 5) _____ with each other in certain circumstances. The gold, lead and carbon are all in the 6) _____ form, while oxygen and nitrogen are in the 7) _____ form. If you raise the temperature high enough, gold or lead may also become 8) _____ . If you burn this carbon, it will be in a gassy 9) _____ . If we can no longer divide a substance into a smaller chunk and it still carries the properties, we call it 10) _____ .

4. *Share your opinions with your partners on the following topics for discussion.*

1) Do you like the lecture from Khan Academy? Why do you enjoy such a lecture? Please summarize the features of MOOCs (Massive Open Online Course) in Khan Academy.

2) Can you use a few lines to list what's your understanding about elements and atoms? Please use an example to clarify your thoughts.

Task Two　Elements and Atoms (Ⅱ)

视频链接及文本

New Words

proton n.[化]质子	quantum n.[物]量子
nucleus n.原子核	bound adj.绑住的
constituent n.构成部分	charge n.电荷
electron n.电子	positive adj.正电的
neutron n.中子	negative adj.负电的
model n.模型	label v.把……列为
abstract adj.抽象的	velocity n.速率,速度
conceptualize v.概念化	interact v.相互作用
version n.版本	configure v.配置;设定
nuance n.细微差别	repel v.击退
buzz v.发嗡嗡声	swipe v.重击
orbit v.环绕轨道运行	affinity n.相吸
planet n.[天]行星	swap v.用……替换

Exercises

1. *Watch the video for the first time and choose the best answers to the following questions.*

1) Which of the following is not inside an atom?

　　A. Proton.　　　　　　　　　　B. Neutron.

　　C. Electron.　　　　　　　　　　D. Protein.

2) _____ carries positive charge inside an atom.

    A. Proton   B. Neutron

    C. Electron   D. Nucleus

3) Which buzzes around the nucleus inside an atom?

    A. Proton.   B. Neutron.

    C. Electron.   D. Nucleus.

4) Why don't the electrons fly off the nucleus? It is because _____.

    A. the electrons carry negative charge

    B. the electrons are too heavy to fly away

    C. the electrons are inside the nucleus

    D. the electrons are bound to the nucleus through a rope

5) Which of the following is true about the symbol ^{12}C?

    A. There are twelve protons in the carbon nucleus.

    B. The total number of protons and neutrons in a carbon nucleus is 12.

    C. There are twelve neutrons in the carbon nucleus.

    D. There are twelve electrons in the carbon nucleus.

2. *Watch the video again and decide whether the following statements are true or false.*

    1) An atom may lose its electron under some circumstances. (　)

    2) The neutron inside the nucleus carries positive charge. (　)

    3) If there are eight protons inside the nucleus, this atom should be a carbon. (　)

    4) The number of protons may be changed while the property of the atom doesn't. (　)

    5) There is only one proton in the atom of hydrogen. (　)

3. *Watch the video for the third time and fill in the following blanks of the table.*

Chemical Symbols	Number of Protons	Number of Neutrons	Number of Electrons	Positive/Negative Charge
He				
C⁺				
N				
O⁻				
F				

4. *Share your opinions with your partners on the following topics for discussion.*

    1) Do you know the composition of an atom? Which particle carries positive charge? Which particle carries no charge? Which particle carries negative charge?

    2) Can you use a metaphor to illustrate such a set of abstract concepts so as to make the audience better understand them?

IV. Talking

Task One　Classical Sentences

Directions：*In this section, some popular sentences are supplied for you to read and to memorize. Then, you are required to simulate and produce your own sentences with reference to the structure.*

General Sentences

1. Where do you live?

   你住在哪里？

2. I live on the Washington Street.

   我住在华盛顿街。

3. I'm Mr. Smith's next-door neighbor.

   我是史密斯先生的邻居。

4. You live here in the city, don't you?

   你住在这个城市，对吗？

5. I'm from out of town.

   我住在城外。

6. How long have you lived here?

   你在这住了多久了？

7. I've lived here for five years.

   我在这住了 5 年了。

8. Where did you grow up?

   你在哪长大？

9. I grew up right here in this neighborhood.

   我就在这附近长大。

10. My friend spent his childhood in California.

    我朋友在加利福尼亚度过了他的童年。

11. He lived in California until he was seventeen.

    他十七岁以前都住在加利福尼亚。

12. There have been a lot of changes here in the last 20 years.

    在过去二十年间，这里发生了很多的变化。

13. There used to be a grocery store on the corner.

    以前拐角处有一个杂货店。

14. All of those houses have been built in the last ten years.

    那些房子都是在最近十年里建成的。

15. They're building a new house up the street.
这个街上正在建一栋新房子。

16. If you buy that house, will you spend the rest of your life there?
如果你买了那栋房子,是不是打算就在那里住到老?

17. Are your neighbors friendly?
你的邻居们友好吗?

18. We all know each other pretty well.
我们都很了解对方。

19. Who bought that new house down the street?
大街那头的那栋新房子被谁买下了?

20. An old man rented the big white house.
一个老人租了那栋白色大房子。

21. We're looking for a house to rent for the summer.
我们在找一栋房子夏天租住。

22. Are you trying to find a furnished house?
你是不是在找一栋带家具的房子?

23. That house is for sale. It has central heating. It's a bargain.
那栋房子在出售,中央供暖,价格很合理。

24. This is an interesting floor plan. Please show me the basement.
这是一个不错的平面图。麻烦带我到地下室去看看。

25. The roof has leaks in it, and the front steps need to be fixed.
屋顶漏水,前面台阶也需要修理。

26. We've got to get a bed and a dresser for the bedroom.
我们得在卧室弄张床和一个梳妆台。

27. They've already turned on the electricity. The house is ready.
房子已经通电,可以入住了。

28. I'm worried about the appearance of the floor.
我对地板的外观甚感担忧。

29. What kind of furniture do you have? Is it traditional?
你这里有什么样式的家具?是传统型的吗?

30. We have drapes for the living room, but we need kitchen curtains.
客厅的窗帘我们有了,但我们需要厨房的窗帘。

31. The house needs painting. It's in bad condition.
这房子得粉刷了,情况很糟。

32. In my opinion, the house isn't worth the price they're asking.
依我看,这间房子根本不值他们要的价钱。

33. Will you please measure this window to see how wide it is?
请你测量一下这个窗口,看看它有多宽?

34. This material feels soft.
    这种材料摸上去很软。

35. —Can you tell me where Peach Street is?
    —Two blocks straight ahead.
    ——你能告诉我皮彻大街在哪吗？
    ——一直朝前走，过两个街区就到了。

36. Should I go this way, or that way?
    我要走这条路还是那条？

37. Go that way for two blocks, and then turn left.
    走那条路，穿过两个街区后向左拐。

38. How far is it to the university?
    到大学还有多远？

39. The school is just around the corner. It's a long way from here.
    学校就在拐角处。从这儿走，还有很长的一段路。

40. Are you married?
    你结婚了吗？

41. No, I'm not married. I'm still single.
    没有，我还是单身。

42. Your niece is engaged, isn't she?
    你的侄女订婚了，是吗？

43. When is your grandparents' wedding anniversary?
    你祖父母的结婚纪念日是什么时候？

44. How long have they been married?
    他们结婚多久了？

45. They've been married for three decades.
    他们结婚三十年了。

46. We're trying to plan our future.
    我们在努力计划我们的未来。

47. I've definitely decided to go to California.
    我已经决定去加利福尼亚了。

48. Who are you writing to?
    你在给谁写信呢？

49. I'm writing to a friend of mine in South America.
    我在给我南美的一位朋友写信。

50. —How long has it been since you've heard from your uncle?
    —I feel guilty because I haven't written to him lately.
    ——从你收到你叔叔的信到现在已经多久了？
    ——我总觉得很内疚，因为我最近没有给他写信。

Specialized Sentences

1. Different substances tend to have different properties.
   不同的物质有不同的属性。

2. One substance may reflect the light in a certain way, while another may not at all.
   一种物质可能以某种方式反射光线，而另外一种物质根本不反射光线。

3. Every substance may be in the state of solid, liquid or gas as long as the environment meets the requirements.
   只要环境满足要求，每种物质可以是固态、液态或气态。

4. This right here is a picture of carbon in the graphite form.
   这一张是以石墨形式存在的碳的图片。

5. If you raise the temperature high enough, metals such as gold and lead may turn into the liquid form.
   如果温度升得足够高，像黄金和铅之类的金属可能会转化成液态。

6. If you burn the carbon, you can get it to a gassy state and you can release it into the air.
   如果你燃烧碳，你能把它变成气态，并可将其释放到大气中。

7. That leads to a natural question that used to be a philosophical question.
   这就会引发出一个自然问题，一个也曾经是哲学问题的问题。

8. Now we can answer the question a little better.
   现在我们能够更好地回答这个问题了。

9. If you keep breaking down that carbon into smaller and smaller chunks, is there some smallest chunk of this substance that still has the properties of carbon?
   如果你把碳分成越来越小的块，有没有仍然保持碳的属性的最小的块呢？

10. You may say that water is an element, and in history, people may refer to water as an element.
    你可能会说水是一个元素，而且在历史上，人们是把水当成元素的。

11. But now we know that water is made up of more basic elements.
    但是现在我们知道水是由更基本的元素构成的。

12. All of our elements are listed here in our periodic table of elements.
    我们所有的元素都列在了这张元素周期表里了。

13. C stands for carbon.
    C 代表碳元素。

14. I'm just going through the elements that are very relevant to humanity.
    我只快速提一下与人类紧密相关的元素。

15. Over time, you will familiarize yourself with all of the elements.
    随着时间的推移，你将对所有的元素非常熟悉。

16. The most basic unit of any of these elements is atom.
    所有这些元素最基本的单位是原子。

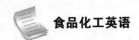

17. So if you were going to keep digging in and keep taking smaller and smaller chunks of it, eventually you would get to a carbon atom.

因此，如果你一直深挖下去，把它分的块越来越小，最后你就会得到一个碳原子。

18. You wouldn't be able to break that down any more.

你将无法将其再分。

19. Just to give you an idea, this is really something I really have trouble imagining—atoms are unbelievably small.

坦白说，我真的很难想象——原子小到难以置信的程度。

20. Also, my hair is made of carbon.

我的头发也是由碳原子构成的。

21. In fact, most of the substance is made of carbon, too.

事实上，大部分的物质都是由碳原子构成的。

22. I would ask you how many carbon atoms wide my hair is.

我想问你，我的头发有多少个碳原子那么宽。

23. If you took a cross-section of my hair, I would like to tell you that there are one million carbon atoms.

如果你得到我头发的横截面，我想告诉你这个横截面的宽度是一百万个碳原子。

24. You could string one million carbon atoms across the width of the average human hair.

你把一百万个碳原子串起来，就能够得到人类头发丝的平均直径。

25. That gives you a sense of how small an atom is.

这会让你感觉到一个原子有多小。

26. Now it would be pretty cool in and of itself to imagine that there would be a million carbon atoms just going along it.

现在，一想到有一百万个原子仅仅是它的直径，这本身就很酷。

27. A gold atom is made up of even more fundamental particles.

一个金原子是由更基础的微粒构成的。

28. They are actually defined by the arrangement of those fundamental particles.

事实上，它们是由这些基本粒子的排列方式来定义的。

29. If you were to change the number of the fundamental particles you have, you could change the properties of this element.

如果你想改变你所拥有的基本微粒的个数，你就会改变这个元素的性质。

30. Or you could change the element itself.

或许你就会改变这个元素本身。

31. Just to understand a little better, let's talk about those fundamental particles.

为了能够更容易理解，我们讨论一下那些基本的微粒。

32. So this is what defines an element—the number of protons in the nucleus.

那么原子核中质子的数目确定了元素。

33. When you look at the periodic table over here, they are actually written in order of atomic number.

当你看这里的元素周期表时,这些元素事实上是按照原子序数的顺序排列的。

34. The atomic number is literally the number of protons in the element.

从字面上讲,原子序数就是元素中质子的数目。

35. So by definition, hydrogen has one proton.

因此,根据定义,氢原子有一个质子。

36. You cannot have carbon of seven protons, and if you did, it could be nitrogen. It would not be carbon any more.

碳原子不能有七个质子;如果有七个质子,它就是氮原子。它就不再是碳原子了。

37. If you somehow add a proton to oxygen, it would be fluorine.

如果你给氧原子增加一个质子,它就变成了氟原子了。

38. For each of these elements in the periodic table, the number of proton is equal to the atomic number.

对于元素周期表上任何一个元素,质子的数目等于原子序数。

39. They put the number up here because that is the defining characteristic of an element.

他们把数字写在上面,因为它是一种元素的概念性的特征。

40. The other two constituents of an atom—I guess we could call it that way— are electron and neutron.

原子的另外两个组成部分,我们可以称作电子和中子。

41. It will get a little bit more abstract and really hard to conceptualize.

这会变得更加抽象,而且很难给它一个概念。

42. Carbon 12, which is a version of carbon, will also have six neutrons.

碳 12 是碳元素的一种,也有 6 个中子。

43. You can have versions of carbon that have a different number of neutrons.

不同种类的碳元素的中子数目是不一样的。

44. So let me draw a carbon 12 nucleus.

那么我来画一个碳 12 的原子核。

45. The reason why we write it carbon 12 is that the total number of protons and neutrons inside of its nucleus is 12.

我们写作碳 12 是因为原子核中质子和中子的数目之和是 12。

46. And this carbon by definition has an atomic number of 6.

根据定义,碳原子的原子序数是 6。

47. So at the center of the carbon atom we have this nucleus.

在碳原子的中心就是我们所说的原子核。

48. If carbon 12 is neutral, it will also have 6 electrons.

如果碳 12 是中性,它也就有 6 个电子。

49. You can imagine that electrons are kind of moving around, buzzing around this nucleus.

你可以想象,电子就像蜜蜂一样在原子核周围移动、转圈。

50. They don't orbit the way that a planet, say, orbits around the sun.

它们不像行星绕着太阳运行一样,按照一定的轨道运动。

Task Two　Sample Dialogue

Directions：*In this section, you are going to read several times the following sample dialogue about the relevant topic. Please pay special attention to five C's (culture, context, coherence, cohesion and critique) in the dialogue and get ready for a smooth communication in the coming task.*

In a chemistry class

(*A teacher and his students are talking about the importance of learning chemistry.*)

Teacher： Good morning, boys and girls. We are going to start a new subject—chemistry. Do you have any idea about chemistry before you come to this class?

Student A： I just know that chemistry is a brother to physics. They have too much knowledge in common.

Teacher： Could you please give an example?

Student A： In physics, charge is a very important part. In chemistry, we are going to touch the same part.

Student B： Also, some abbreviations are universal in both subjects. We use the letter t to stand for temperature, and we use the letter v to stand for speed. And so on.

Student A： Chemistry has another brother—biology.

Teacher： It's a new idea. Please go on talking.

Student A： When we are learning biology, we have done a lot of experiments with tubes and chemicals. From what you carry into the classroom, we can see that this new subject—chemistry —will also be involved with tubes.

Teacher： It's incredibly creative. Does chemistry have any other brothers?

Student B： I think math is also a close brother to chemistry.

Teacher： What are the reasons for such a conclusion?

Student B： Both of them will require very complicated calculation and equations. You know, I am not good at calculation because my mother says I am too careless.

Teacher： To my amazement, chemistry has so many brothers, namely biology, physics and math. From such a discussion, we can have a rough idea about this new subject. Thank you for your sparkling idea and active participation.

Task Three　Simulation and Reproduction

Directions：*The class will be divided into three major groups, each of which will be assigned a topic. In each group, some students may be the teacher, while others may be students. In the process* *of discussion , please observe the principles of cooperation , politeness and choice of words. One of the groups will be chosen to demonstrate the discussion to the class.*

1) Chemistry in our daily life.

2) A funny story related to chemistry in my childhood.

3) The importance of learning chemistry.

Task Four   Discussion and Debate.

Directions: *The class will be divided into two groups. Please choose your stand in regard to the following controversy and support your opinions with scientific evidences. Please refer to the specialized terms and classical sentences in the previous parts of this unit.*

In the traditional Chinese culture , it was believed that everything in the universe consists of five elements , namely *Jin* , *Mu* , *Shui* , *Huo* and *Tu*. However , the modern chemistry holds the opinions that there are more than 100 chemical elements and it seems that the family of elements is getting larger and larger. Which party do you agree with? Why?

V. After-class Exercises

1. *Match the English words in Column A with the Chinese meaning in Column B.*

A	B
1) atom	a. 原子核
2) element	b. 质子
3) proton	c. 中子
4) nucleus	d. 电子
5) electron	e. 元素
6) neutron	f. 碳
7) carbon	g. 氧
8) oxygen	h. 铅
9) lead	i. 物质
10) substance	j. 原子

2. *Fill in the following blanks with the words or phrases in the word bank. Some may be chosen more than once while some may not be elected. Change the forms if it's necessary.*

oxygen	proton	atom	velocity	electrons	liquid
nucleus	negative	solid	affinity	hydrogen	graphic
periodic table	interact	eighteen	neutron		

1) The smallest chunk that keeps the properties in a substance is a(n) _____ .

2）Water is made up of two elements：_____ and _____.

3）In a nucleus, a _____ carries positive charge while an electron carries _____ charge.

4）Inside the atom of hydrogen, there is no _____ but one proton.

5）In the chemical symbol O^{2-}, two _____ are added to the oxygen atom.

6）There are three states in nature, including _____, _____ and gas.

7）The electrons are buzzing around the _____ in an atom.

8）Inside an atom, the number of _____ is equal to the atomic number.

9）In the chemical symbol Ca^{2+}, there are 20 protons and _____ electrons.

10）In the _____, the elements are listed in order of atomic numbers.

3. *The following is a periodic table with some elements missing. Can you help the missing members go back home?*

	I A																	0
1	H	IIA											IIIA	IVA	V A	VIA	VIIA	He
2		Be											B	C	N		F	Ne
3		Mg	IIIB	IVB	VB	VIB	VIIB		VIIIB		I B	IIB						
4		Ca	Sc	Ti	V	Cr	Mn	Fe	Co	Ni	Cu	Zn	Ga	Ge	As		Br	Kr
5		Sr	Y	Zr	Nb	Mo	Tc	Ru	Rh	Pd	Ag	Cd	In	Sn	Sb		I	Xe
6		Ba	La	Hf	Ta	W	Re	Os	Ir	Pt	Au	Hg	Tl	Pb	Bi	Po	At	Rn
7	Fr	Ra	Ac	Rf	Db	Sg	Bh	Hs	Mt	Ds	Rg	Cn	Uut	Fl	Uup			

镧系	La	Ce	Pr	Nd	Pm	Sm	Eu	Gd	Tb	Dy	Ho	Er	Tm	Yb	Lu
锕系	Ac	Th	Pa	U	Np	Pu	Am	Cm	Bk	Cf	Es	Fm	Md	No	Lr

4. *Translate the following sentences into English.*

1）如果温度降得足够低，像氢气和氧气之类的气体可能会转化成液态。

_____

2）如果你把铅分成越来越小的块，有没有仍然保持铅的属性的最小的块呢？

_____

3）我们所有的元素都按照原子序号，排列在了这张元素周期表里了。

_____

4）事实上，它们是由质子、中子和电子这些基本粒子的排列方式来定义的。

_____

_____

5) 坦白说,我真的很难想象原子核中竟然包含有更小的质子和中子。

_____

_____

5. *Please write an essay of about 120 words on the topic*"**Application of Chemistry in Our Life**". *Some specific examples will be highly appreciated and watch out the spelling of some specialized terms you have learnt in this unit.*

_____

_____

_____

_____

_____

_____

_____

_____

_____

_____

_____

_____

_____

_____

_____

_____

VI. Additional Reading

Brief Introduction on the History of BASF

BASF SE is a German chemical company and the largest chemical producer in the world. The BASF Group comprises(构成) subsidiaries(子公司) and joint ventures(合资公司) in more than 80 countries and operates six integrated production sites(综合生产基地) and 390 other production sites in Europe, Asia, Australia, the Americas and Africa. Its headquarters(总部) is located in Ludwigshafen, Germany. BASF has customers in over 190 countries and supplies products to a wide variety of industries. Despite its size and global presence, BASF has received relatively little public attention since it abandoned manufacturing and selling BASF-branded consumer electronics products in the 1990s.

At the end of 2015, the company employed more than 122 000 people, with over 52 800 in Germany alone. In 2015, BASF posted sales of €70.4 billion and income from operations (营业) before special items of about €6.7 billion. The company is currently expanding its international activities with a particular focus on Asia. Between 1990 and 2005, the company invested €5.6 billion in Asia, for example in sites near Nanjing and Shanghai, China and Mangalore in India.

BASF is listed on the Frankfurt Stock Exchange (法兰克福证券交易所), London Stock Exchange, and Zurich Stock Exchange. The company delisted (退市) its ADR (American depositary receipt 美国存托凭证) from the New York Stock Exchange in September 2007. The company is a component (组成部分) of the Euro Stoxx 50 stock market index (欧洲斯托克 50 股票市场指数).

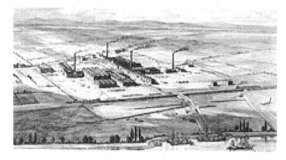

BASF (Badische Anilin und Soda Fabrik, or, in English, Baden Aniline and Soda Factory) was founded on 6 April 1865 in Mannheim, in the German-speaking country of Baden by Friedrich Engelhorn. It had been responsible for setting up a gasworks (煤气厂) and street lighting for the town council (市政厅) in 1861. The gasworks produced tar (焦油) as a byproduct (副产品), and Engelhorn used this for the production of dyes. BASF was set up in 1865 to produce other chemicals necessary for dye production, notably soda and acids (酸). The plant, however, was erected (建立) on the other side of the Rhine river at Ludwigshafen because the town council of Mannheim was afraid that the air pollution of the chemical plant could bother the inhabitants (居民) of the town. In 1866 the dye production processes were also moved to the BASF site.

Dyes

The discovery in 1857 by William Henry Perkin that aniline (苯胺) could be used to make intense (浓烈的) colouring agents (着色剂) had led to the commercial production of synthetic dyes (合成染料) in England from aniline extracted (提炼) from coal tar. BASF recruited (招收, 雇佣) Heinrich

Caro, a German chemist with experience of the dyestuff (染料) industry in England, to be the first head of research. Caro developed a synthesis (合成) for alizarin (茜素) [a natural pigment (色素) in madder (茜草)], and applied for a British patent (专利) on 25 June 1869. Coincidentally (巧合地), Perkin applied for a virtually identical (完全一样的) patent on 26 June 1869, and the two companies came to a mutual commercial agreement about the process.

Further patents were granted for the synthesis of methylene (亚甲基) blue and eosin (曙红), and in 1880 research began to try to find a synthetic process for indigo (靛蓝) dye, though this was not successfully brought to the market until 1897. In 1901, some 80% of the BASF production was dyestuffs.

BASF main laboratory in Ludwigshafen, 1887

Soda

Sodium carbonate (碳酸钠) (soda) was produced by the Leblanc process until 1880 when the much cheaper Solvay process became available. BASF ceased (停止) to make its own and bought it from the Solvay company thereafter.

Sulfuric acid (硫酸)

Sulfuric acid was initially (最初) produced by the lead chamber process (铅室法), but in 1890 a unit using the contact process (接触法) was brought on stream, producing the acid at higher concentration (浓度) (98% instead of 80%), and at a lower cost. This followed extensive research and development by Rudolf Knietsch, for which he received the Liebig Medalin 1904.

Ammonia (氨)

The development of the Haber process from 1908 to 1912 made it possible to synthesize (合成) ammonia (a major industrial chemical as the primary source of nitrogen), and, after acquiring exclusive rights to the process, in 1913 BASF started a new production plant in Oppau, adding fertilizers (肥料) to its product range. BASF also acquired and began mining (采矿) anhydrite (硬石膏) for gypsum (石膏) at the Kohnstein in 1917.

IG Farben

In 1916, BASF started operations at a new site in Leuna, where explosives were produced during the First World War. On 21 September 1921, an explosion occurred in Oppau, killing 565 people. The Oppau explosion was the biggest industrial accident in German history. Under the leadership of

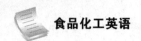

Carl Bosch, BASF founded IG Farben with Hoechst, Bayer, and three other companies, thus losing its independence. BASF was the nominal(名义上的) survivor, as all shares were exchanged for BASF shares prior to the merger(合并). Rubber, fuels, and coatings(涂层) were added to the range of products.

World War II

Following the appointment of Adolf Hitler as Chancellor(总理) in 1933, IG Farben cooperated with the Nazi regime(政权), profiting from guaranteed volumes(数量) and prices, and from the slave labor provided by the government's Nazi concentration camps(集中营). IG Farben also achieved notoriety(臭名昭著) owing to its production of Zyklon-B(齐克隆B), the lethal(致死的) gas used in Nazi extermination camps(灭绝集中营). In 1935, IG Farben and AEG presented the magnetophone(磁电子带录音机)—the first tape recorder—at the Radio Exhibition in Berlin.

The Ludwigshafen site was almost completely destroyed during the Second World War and was subsequently(随后) rebuilt. The allies (盟军) dissolved(解散) IG Farben in November 1945. Both the Ludwigshafen and Oppau plants were of strategic importance for the war because the German military needed many of their products, e.g. synthetic rubber(合成橡胶) and gasoline. As a result, they were major targets for air raids(空袭). Over the course of the war, Allied bombers(轰炸机) attacked the plants 65 times.

Shelling(炮轰) took place from the autumn of 1943 and saturation bombing(饱和轰炸) inflicted(造成) extensive damage. Production virtually(几乎) stopped by the end of 1944.

Due to a shortage of male workers during the war, women were conscripted(招收) to work in the factories, and later prisoners of war and foreign civilians (平民). Concentration camp inmates(犯人) did not work at the Ludwigshafen and Oppau plants.

In July 1945, the American military administration confiscated(没收, 充公) the entire assets (资产) of IG Farben. That same year, the Allied Commission(盟军委员会) decreed (裁定, 决定) that IG Farben should be dissolved. The sites at Ludwigshafen and Oppau were controlled by French authorities.

BASF refounded(重新建立)

On 28 July 1948, an explosion occurred at a BASF site in Ludwigshafen, killing 207 people and injuring 3818. In 1952, BASF was refounded under its own name following the efforts of Carl Wurster. With the German economic miracle in the 1950s, BASF added synthetics such as nylon(尼龙) to its product range. BASF developed polystyrene(聚苯乙烯) in the 1930s and invented Styropor in 1951.

Production abroad

In the 1960s, production abroad was expanded and plants were built in Argentina(阿根廷), Australia, Belgium(比利时), Brazil, France, the United Kingdom, India, Italy, Japan, Mexico,

Spain and the United States. Following a change in corporate(公司的) strategy in 1965, greater emphasis was placed on higher-value products such as coatings, pharmaceuticals(制药), pesticides(杀虫剂) and fertilizers. Following German reunification(统一), BASF acquired a site in Schwarzheide, eastern Germany, on 25 October 1990. It expanded to Podolsk, Russia, in 2012, and to Kazan in 2013.

Takeovers(接管)

In 1968, BASF (together with Bayer AG) bought the German coatings company Herbol. BASF completely took over the Herbol branches in Cologne and Würzburg in 1970. Under new management, the renewal(更新) and expansion of the trademark(商标) continued. After an extensive reorganization and an increasing international orientation(指向) of the coatings business, Herbol became part of the new founded Deco GmbH in 1997.

BASF bought the Wyandotte Chemical Company, and its Geismar, Louisiana chemical plant in the early 1980s. The plant produced plastics, herbicides(除草剂) and antifreeze(防冻剂). BASF soon tried to operate union-free(没有工会), having already reduced or eliminated union membership in several other US plants. Challenging the Geismar OCAW union resulted in a labor dispute(争端) that saw members locked out from 1984 to 1989 and eventually winning their case. A worker solidarity committee(工人团结委员会) at BASF's headquarters plant in Ludwigshafen, Germany, took donations(捐赠) from German workers to support the American strikers and organized rallies(集会) and publicity(宣传) in support. The dispute was the subject of an academic study. The union also exposed major accidental releases of phosgene(碳酰氯), toluene(甲苯) and other toxic gases, these being publicized in the local media and through a video, *Out of Control*. A court threw out a $66 700 fine against BASF for five environmental violations(破坏) as "too small".

BASF's European coatings business was taken over by AkzoNobelin 1999. BASF bought the Engelhard Corporation for $4.8 billion in 2006. Other acquisitions(收购) in 2006 were the purchase of Johnson Polymer and the construction chemicals business of Degussa.

The acquisition of Johnson Polymer was completed on 1 July 2006. The purchase price was $470 million on a cash and debt-free basis. It provided BASF with a range of water-based resins(树脂,松香) that complements its portfolio(证券投资组合) of high solids and UV resins for the coatings and paints industry and strengthened the company's market presence, particularly in North America.

The acquisition of Degussa AG's construction chemicals business was completed in 2006. The

BASF Portsmouth Site in the West Norfolk area of Portsmouth, Virginia, United States
The plant is served by the Commonwealth Railway(联邦铁路公司)

21

purchase price for equity (股本) was about €2.2 billion. In addition, the transaction was associated with debt of €500 million. The company agreed to acquire Ciba [formerly(以前) part of Ciba-Geigy] in September 2008. The proposed deal was reviewed by the European Commissioner for Competition(欧洲竞争委员). On 9 April 2009, the acquisition was officially completed.

On 19 December 2008, BASF acquired U. S.-based Whitmire Micro-Gen together with U. K.-based Sorex Ltd, Widnes, Great Britain. Sorex is a manufacturer of branded chemical and non-chemical products for professional pest(疫情) management. In March 2007 Sorex was put up for sale with a price tag of about £100 million.

In May 2015, BASF agreed to sell parts of its pharmaceutical ingredients(部分) business to Swiss drug manufacturer Siegfried Holding for a fee of €270 million, including assumed debt.

In October 2017, BASF announced it would buy seed and herbicide businesses from Bayer for €5.9 billion ($7 billion), as part of its acquisition of Monsanto.

(*If you want to find more information about this corporation, please log on https://en.wikipedia. org/wiki/BASF*)

1. *Read the passage quickly by using the skills of skimming and scanning. And choose the best answer to the following questions.*

   1) Why does BASF receive relatively little public attention despite its size and global presence?

   A. Because it abandoned manufacturing and selling BASF-branded consumer electronics products in the 1990s.

   B. Because it was merged by another big corporation in the 1990s.

   C. Because the biggest industrial accident happened in one of its factories in the 1990s.

   D. Because it went bankrupt in the 1990s.

   2) At the end of 2015, the company employed more than _____ people in the world.

   A. 52 800   B. 122 000

   C. 704 000   D. 67 000

   3) In 2015, BASF posted sales of _____.

   A. €5.6 billion   B. €6.7 billion

   C. €70.4 billion   D. €7.5 billion

   4) Besides Nanjing, the company built its sites in _____ in China between 1990 and 2005.

   A. Xi'an   B. Chengdu

   C. Wuhan   D. Shanghai

   5) At present, BASF is listed in all of the following stock exchanges except _____.

   A. Frankfurt Stock Exchange   B. New York Stock Exchange

   C. London Stock Exchange   D. Zurich Stock Exchange

6) When was BASF founded in Mannheim, in the German-speaking country of Baden by Friedrich Engelhorn?

    A. On 6 April 1865.   B. On 6 April 1861.

    C. On 6 April 1866.   D. On 6 April 1870.

7) In 1857, William Henry Perkin discovered that _____ could be used to make intense coloring agents.

    A. alizarin   B. methylene

    C. indigo   D. aniline

8) In 1916, BASF started operations at a new site in Leuna, where _____ were produced during the First World War.

    A. paints   B. herbicides

    C. explosives   D. pesticides

9) Following a change in corporate strategy in 1965, greater emphasis was placed on higher-value products including all of the following except _____ .

    A. coatings   B. pharmaceuticals

    C. fertilizers   D. explosives

10) In October 2017, BASF announced it would buy _____ businesses from Bayer for €5.9 billion ($7 billion), as part of its acquisition of Monsanto.

    A. seed and herbicide   B. washing powder

    C. paint and pesticide   D. pesticides and fertilizers

2. *In this part, the students are required to make an oral presentation on either of the following topics.*

    1) The secrets of BASF's success.

    2) The lessons from BASF's development history.

习题答案

Unit Two Nutrients

I. Pre-class Activity

Directions: *Please read the general introduction about Günter Blobel and have a discussion about the important qualities of being a great scientist with your classmates.* (*https://www.questia.com/library/journal/1G1-82007569/cell-biologist-dr-gunter-blobel-nobel-laureate-on*)

Günter Blobel

Günter Blobel (21 May 1936 – 18 February 2018) was the winner of the Nobel Prize in Physiology or Medicine in 1999. Dr. Blobel was awarded the prize for his discovery that "proteins have intrinsic signals that governs their transport and localization in the cell", the Swedish Karolinska Institute noted in its citation.

Dr. Blobel's research built on earlier investigations into how the cell works. Former researchers then didn't know how the cell worked or how large proteins found their way into the organelles through the tightly sealed membranes. In 1971, Dr. Blobel and his colleague proposed a daring idea called the "signal hypothesis". It theorized that "proteins secreted out of a cell contain an intrinsic signal that governs them to, and across, membranes".

It took Dr. Blobel and his research team nearly 30 years to find that intrinsic signal. Eventually, Dr. Blobel found that each newly made protein has an organelle-specific address, a stretch of the protein called a signal sequence, which is recognized by receptors on an organelle's surface. The binding of the signal sequence to its receptor opens a watery channel of the membrane through which the protein can travel. The signal sequence has been dubbed the "address tag", or ZIP code, that helps proteins find their correct locations within the cell.

The finding is important because it helps explain the molecular mechanism behind genetic diseases like cystic fibrosis, which is caused when proteins fail to reach their proper destination.

II. Specialized Terms

Directions : *Please memorize the following specialized terms before the class so that you will be able to better cope with the coming tasks.*

adipose tissue 脂肪组织

alga n.海藻;藻类

aroma n.香味,芳香;风格,风味

aromatic adj.芳香的,有香气的

build up 增进,增强

carbohydrate n.碳水化合物

carotene n.胡萝卜素(等于 carotin)

casein n.酪蛋白

catsup n.番茄酱

cellulose n.纤维素

cholesterol n.胆固醇

coagulated protein 凝固蛋白质

descendant n.子系体,后代

eased adj.膨松的

edible adj.可食用的

elastin.弹性蛋白

enterotoxin n.肠毒素

enzyme n.酶

epinephrine n.肾上腺素

estrogen n.雌激素

fiber n.纤维

film n.薄膜(层)

fodder n.材料

folic acid 叶酸

fragrance n.香味

fruitarian n.果食者

glucose n.葡萄糖

hemoglobin n.血红蛋白

hemolytic toxin 溶血毒素

hemosiderin n.血铁黄素

high-pulp 高果肉

inert adj.惰性的

laver n.紫菜

leaf protein 叶蛋白

legume n.豆科植物

lignin n.木质素

lime n.石灰;酸橙

lipid n.脂质

macroelement n.宏量元素

marine food 海产食品,海味

matrix n.基质

melanin n.黑色素

metabolism n.新陈代谢

mitochondrion n.线粒体

mollusc n.软体动物

muscle meat 瘦肉

Neolithic adj.新石器时代的

nerve n.神经

niacin n.尼克酸

nori n.海苔片;紫菜

nourishment n.食物,滋养品

amino acid 氨基酸

nutrient n.营养物,滋养物

nutrition n.营养;营养学

nutritional adj.营养的,滋养的

odor n.气味

opaque adj.不透明的

orchard n.果园

originate v.起源于

output n.产量

ovum n.卵细胞

pastry n.焙烤(面制)食品,发面点心,面制
  糕点

peanut kernel 花生仁

pectic enzyme 果胶酶

pectin n.果胶

25

peptide n.肽

perception n.感知;觉察

pericarp n.果皮

photosynthesis n.光合作用

physiological adj.生理的

pickle n.泡菜,咸菜

pie n.馅饼

pod n.豆荚

portion n.一部分,一份

preserved food 罐头食品

probiotic adj.前生命期的

protease n.蛋白酶

protein n.蛋白质

protein isolate 分离蛋白

prune n.西梅干

pulp n.果肉

quick-cooking food 速煮食品

quick-frozen food 速冻食品

quick-serve meal 快餐食品

raisin n. [常用复数]葡萄干

reductase n.还原酶

reserves n.储量,储备

retinol n.视黄醇

riboflavin n.核黄素

ribosome n.核糖体

satiety n.饱腹感

season v.调味

seaweed n.海草,海藻

sesame n.芝麻

sesame oil 芝麻油,香油

set milk 凝乳食品

simultaneous adj.同时发生的

sitology n.食品学,营养学

skeletal muscle 骨骼肌

skillet n. [英]长柄(矮脚)小锅;[美]平底锅(= frying pan)

III. Watching and Listening

Task One　Carbohydrate and Energy

视频链接及文本

New Words

sustained adj.持久的,持续的

genetically adv.遗传学方面的,基因方面的

identical adj.同一的;同样的 n. 完全相同的事物;同卵双胞胎

sporty adj.像运动家的

military adj.军事的;军用的

grueling adj. 紧张的,激烈的,使极度疲劳的

split v.分裂;分开

cereal n.谷类植物

molecule n.分子

strenuous adj.费力的;紧张的

forge ahead 稳步前进

pasta n.面团(用以制意大利通心粉、细面条等),意大利面食

tamper v.篡改 ;(用不正当手段)影响,干预

accessible adj.易接近的;可理解的

collapse v.倒塌;崩溃

morale n.士气;精神面貌

endurance n.忍耐,忍耐力

Exercises

1. *Watch the video for the first time and choose the best answers to the following questions.*

1) Which of the following food is not rich in carbohydrates?

    A. Pasta.   B. Rice.

    C. Tomato.   D. Bread.

2) The purpose of the experiment is to _____.

    A. prove the existence of carbohydrate

    B. examine whether a high-carb diet really keeps people going for longer

    C. find out the limited function of carbohydrates

    D. prove carbohydrate is more important than water

3) The four sets of identical sporty twins come from around the globe except _____.

    A. France   B. German

    C. England   D. America

4) The twins in this experiment share the same conditions except _____.

    A. genes   B. strong will to win

    C. diet   D. sporty figure

5) If you're working out physically all day, _____ will be an ideal diet for your performance.

    A. a high-carb diet   B. much protein

    C. vitamin-rich food   D. dietary fiber

2. *Watch the video again and decide whether the following statements are true or false. Write T or F in the corresponding brackets.*

1) The twins choose their teammates by themselves.(　　)

2) The red team have carb-filled feast of cereals, sugar and toast as the breakfast.(　　)

3) During the first two hours, the performances of the two teams seem to be working well. (　　)

4) The high-carb blue team has a much better performance than the red in the cycling.(　　)

5) Fat and protein diet provide the red team with the right energy to get them catch up with the blue team.(　　)

3. *Watch the video for the third time and fill in the following blanks.*

    Team at the rear is now red team. Team in the front is now blue team. Each twin will be 1)_____ against their genetic 2)_____ image. They don't know yet, but tomorrow will be the most grueling day of their lives.

    Next morning, the two teams split up for breakfast. They're eating 3)_____, so they won't find out that we're about to tamper with their diets. The blues are going to eat a high carb diet. So breakfast is a carb-filled 4)_____ of cereals, sugar and 5)_____. Carbs are packed full of molecules that the body can turn into 6)_____ really easily during strenuous 7)_____.

    The red team are going to eat a low carb diet—eggs and bacon, a packed full of 8)_____, and more importantly fat, but no carbs. Fat molecules are 9)_____ in a way

that the body finds much 10)_____ to turn into energy during strenuous exercise.

4. *Share your opinions with your partners on the following topics for discussion.*

1) Who do you think will benefit from the findings of this experiment?

2) How do you believe in this experiment? Is it perfectly reasonable and convincing?

Task Two   Lutein and Macular Degeneration

视频链接及文本

New Words

intensity n.强度;烈度

macular adj.有斑点的,有污点的

degeneration n.退化;恶化

macula n.斑点;斑疹

retina n.[解]视网膜

receptor n.[生]感受器,受体

rod n.杆,拉杆

cone n.圆锥体

UV light 紫外线

pigment n.颜料,色料;[生]色素

compound n.场地;复合物

lutein n.叶黄素,黄体制剂

hypothesis n. 假设;[逻]前提

sprinkling n.少量,一点儿

Exercises

1. *Watch the video for the first time and choose the best answers to the following questions.*

1) In some people,_____ can lead to a form of blindness called macular degeneration.

    A. dim light   B. bacteria

    C. intense sunlight   D. insufficient sleep

2) The macula is _____ in the middle of the retina.

    A. at the back of the eye   B. at the front of the eye

    C. at the middle of the eye   D. at the surface of the eye

3) Within the macula,there are millions of _____.

    A. UV light   B. photo receptors

    C. blue light   D. corneas

4) _____ is asked to test whether spinach is one of the best sources of lutein.

    A. University of Manchester   B. Imperial College London

    C. Cambridge University   D. Oxford

5) What is not recommended by the author to those who want to protect the eyesight?

    A. Kiwi fruit.   B. Egg yolks.

    C. Spinach.   D. Ginger.

2. *Watch the video again and decide whether the following statements are true or false.*

1) Raphel,as well as Jones,never worries about the health of their eyes.(　　)

2) If we could increase the amount of lutein,we should be able to protect people from macular degeneration.(　　)

3) In the United States,the population on average is aging,which means the number of macular degeneration is set to increase.(　　)

4）After eating spinach regularly, Jones does not notice anything encouraging about her eyesight.（　）

5）Without seeing the doctor, Jones is quite sure the new glasses, but not the spinach, is improving her eye condition.（　）

3. *Watch the video for the third time and fill in the following blanks.*

In some people, 1）_____ sunlight can lead to a form of 2）_____ called macular degeneration. This condition results in a severe loss of detail and 3）_____ vision. "In the United States, it is 4）_____ that there are somewhere between 5）_____ million people with age-related macular degeneration. And our 6）_____ on average is aging", which means this number is set to 7）_____.

So to understand macular degeneration, we need to look inside the eye. The macula is at the back of the eye in the middle of the 8）_____. Within the macula, there are millions of photo 9）_____. And these rods and cones are very easily 10）_____ by sunlight, particularly blue and UV light.

4. *Share your opinions with your partners on the following topics for discussion.*

1）Are there any other factors that may influence our eyesight in daily life?

2）How can people take precautions against the damages to our eyes?

IV. Talking

Task One　Classical Sentences

Directions：*In this section, some popular sentences are supplied for you to read and to memorize. Then, you are required to simulate and produce your own sentences with reference to the structure.*

General Sentences

1. What's your nationality? Are you Chinese?
   你是哪国人? 是中国人吗?

2. What part of the world do you come from?
   你来自哪里?

3. I was born in Spain, but I'm a citizen of France.
   我出生在西班牙, 但我是法国公民。

4. Do you know what the population of Japan is?
   你知道日本有多少人口吗?

5. What's the area of the Congo in square miles?
   刚果的国土面积是多少平方英里?

6. Who is the governor of this state?
   谁是该州的州长?

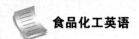

7. According to the latest census, our population has increased.
根据最新的人口普查,我们的人口增加了。

8. Politically, the country is divided into fifty states.
该国从行政上被划分为 50 个州。

9. The industrial area is centered largely in the north.
工业区大部分集中在北方。

10. The country is rich in natural resources. It has large quantities of mineral deposits.
该国自然资源丰富,有大量的矿藏。

11. This nation is noted for its economic stability.
该国以经济稳定而出名。

12. The U.S. is by far the biggest industrial country in the world.
到目前为止,美国是世界上最大的工业国。

13. My home is in the capital. It's a cosmopolitan city.
我的家在首都,它是一个国际大都会。

14. Geographically, this country is located in the southern hemisphere.
从地理位置上讲,这个国家位于南半球。

15. Britain is an island country surrounded by the sea.
英国是一个被海洋环抱的岛国。

16. It's a beautiful country with many large lakes.
这是一个有着若干大湖的美丽国度。

17. The River Thames is the second longest and most important river in Britain.
泰晤士河是英国第二大河,也是英国最重要的河。

18. This part of the country is very mountainous.
这个国家的这部分由众多的山脉覆盖。

19. The land in this region is dry and parched.
这个地区的土地干旱且贫瘠。

20. Along the northern coast there are many high cliffs.
北海岸多危壁断崖。

21. There are forests here, and lumbering is important.
此地多森林,故以伐木业为重。

22. In Brazil, many ancient forests are very well preserved.
在巴西,古老的森林保存十分完好。

23. The scenery is beautiful in the small islands in the Pacific Ocean.
太平洋上一些小岛的景色十分优美。

24. This mountain range has many high peaks and deep canyons.
此山高峰深谷众多。

25. What kind of climate do you have? Is it mild?
你们那里气候怎么样? 温和吗?

26. Britain has a maritime climate—winters are not too cold and summers are not too hot.
    英国属于海洋性气候,冬季不过于寒冷,夏季不过于炎热。

27. Mount Tai is situated in the Western Shandong Province.
    泰山位于山东省西部。

28. How far is it from the shore of the Atlantic to the mountains?
    从大西洋海岸到山区有多远?

29. Lumbering is very important in some underdeveloped countries.
    在一些不发达的国家,伐木业十分重要。

30. What's the longest river in the United States?
    美国最长的河是什么河?

31. Are most of the lakes located in the north central region?
    大部分湖泊是不是在北部的中心地区?

32. As you travel westward, does the land get higher?
    你去西部旅行时,是不是地势越来越高?

33. The weather is warm and sunny here. Do you get much rain?
    这里的天气温暖而晴朗。雨多吗?

34. Because of the warm and sunny weather, oranges grow very well here.
    因为这里气候温暖,光照充足,橘子长势很好。

35. In this flat country, people grow wheat and corn and raise cattle.
    这个国家地势平坦,人们种植小麦、玉米,饲养牲畜。

36. The ground around here is stony and not very good for farming.
    这周围的土地多石,不适合耕种。

37. Is the coastal plain good for farming?
    这种海边的平原有利于发展农业吗?

38. Is the plain along the river good for farming?
    河畔的平原易于发展农业吗?

39. What are the principal farm products in this region?
    这个地区的主要农产品是什么?

40. Milk, butter, and cheese are shipped here from the dairy farms.
    牛奶、黄油、奶酪都从奶牛场运到这里。

41. They had to cut down a lot of trees to make room for farms.
    他们不得不砍伐一些树木,从而为农场提供足够的空间。

42. At this time of the year farmers plow their fields.
    一年中的这个时候农民们会耕种自己的土地。

43. On many farms you'll find cows and chickens.
    在许多农场你都能发现奶牛和鸡。

44. If you have cows, you have to get up early to do the milking.
    如果你有奶牛,你得早起挤牛奶。

45. Tractors have revolutionized farming.
拖拉机使农业发生了革命性的变化。

46. In the United States, there are many factories for making cloth.
美国有很多制布厂。

47. Factories employ both male and female workers.
工厂既雇佣男工,也雇佣女工。

48. If you work in a factory, you usually have to punch a clock.
如果你在工厂工作的话,你就得打卡。

49. Is meat packing a big industry in your country?
肉类包装在你们国家是不是一个大的产业?

50. Is it true that manufacturing of automobiles is a major industry?
汽车制造业是主打产业,是吗?

Specialized Sentences

1. Fats not only yield energy but they are also necessary for the absorption of certain vitamins.
脂肪不仅能够产生能量,而且可以吸收某些维生素。

2. Fats are an essential part of every diet for the production of hormones.
脂肪是所有饮食的必要组成部分,因为它能够产生一些激素。

3. Fats can be divided into saturated fats, monounsaturated fats and polyunsaturated fats.
脂肪分为饱和脂肪、单不饱和脂肪和多不饱和脂肪。

4. All oils should not be considered the same relative to their health effects.
各种油脂对人体健康的作用是不一样的。

5. Saturated fats are typically found as solids at room temperature.
饱和脂肪一般在室温下是固态。

6. Red meat and dairy products, such as cheese and ice cream, contain saturated fats.
红肉和奶制品,例如奶酪和冰激凌中含有饱和脂肪。

7. Fats also play a role in protecting vital organs.
脂肪还有助于保护生命器官。

8. Polyunsaturated fats, as one group of unsaturated fats, are generalized as the good fats.
多不饱和脂肪作为一种不饱和脂肪,一般统称为有益脂肪。

9. Monounsaturated fats are more beneficial than polyunsaturated fats.
单不饱和脂肪比多不饱和脂肪更有益处。

10. Polyunsaturated fats come in 2 main classes: omega-6 and omega-3.
多不饱和脂肪主要有两类: ω-6 和 ω-3。

11. These terms are derived from their chemical structures, and their biological functions.
这些命名来源于它们的化学结构,也源于它们的生物学功能。

12. Americans have increased consumption of omega-6 fats.
美国人已经提高了 ω-6 脂肪的摄入量。

13. Protein is needed in the body for many reasons, most commonly for growth as well as tissue replacement and maintenance.

身体需要蛋白质的原因很多,最常见的是为了促进成长以及替换和维护组织。

14. Proteins are made up of two types of amino acids: essential and non-essential.

蛋白质由两类氨基酸组成:必需氨基酸和非必需氨基酸。

15. True to its names, non-essential amino acids can be synthesized by your body and therefore it is not essential to get them from your daily diet.

正如其名,非必需氨基酸可以由身体本身合成,因此不必通过日常膳食摄入。

16. The other type of amino acids is essential, which means it is necessary to get them from your diet.

另一种氨基酸是必需氨基酸,意味着必须从饮食中摄入。

17. The categories of proteins include complete, incomplete and complementary.

蛋白质包括完全蛋白质、不完全蛋白质和互补蛋白质。

18. A protein is considered complete when it contains all of the essential amino acids.

如果一种蛋白质含有所有的必需氨基酸,那么它属于完全蛋白质。

19. The common sources of complete proteins are animal sources such as meat and milk.

完全蛋白质通常可以从动物身上获取,例如肉类和奶制品。

20. Incomplete proteins are proteins that lack one or more of the essential amino acids.

不完全蛋白质就是那些缺少一种或多种必需氨基酸的蛋白质。

21. Common sources of incomplete proteins are most vegetable proteins.

这类蛋白质通常来源于蔬菜蛋白质。

22. Complementary proteins consist of two or more incomplete proteins that when mixed together, they form a complete protein, such as rice and beans.

互补蛋白质由两种或多种不完全蛋白质组成,当两种不完全蛋白质混合在一起时,可以合成一种完全蛋白质,比如大米和豆类。

23. The vitamins that are carried in foods with fat are the so-called fat-soluble vitamins.

食物脂肪中包含的维生素就是所谓的脂溶性维生素。

24. Fat-soluble vitamins include vitamin D, vitamin E, vitamin K and vitamin A.

脂溶性维生素分为维生素 D、维生素 E、维生素 K 和维生素 A。

25. Because our bodies store excess fat-soluble vitamins in the liver and in fat cell, we don't need to consume them every day to avoid deficiencies.

由于在我们身体里的肝脏和脂肪细胞中贮存着多余的脂溶性维生素,所以我们不必每天摄入。

26. There are high risks of toxicity if too much of a fat-soluble vitamin is ingested.

如果吸收太多某种脂溶性维生素,就有很高的中毒危险。

27. An upper tolerable limit has been defined for each fat-soluble vitamin.

每种脂溶性维生素都有其摄入的上限。

28. Vitamin A is important for eye function, skin health, and may have a role in cancer

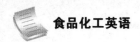

prevention.

维生素 A 对于视觉功能、皮肤健康很重要,也可能有助于预防癌症。

29. Vitamin A also has other roles in the body including promoting teeth and bone development, and as a regulator of cell division and differentiation.

维生素 A 在身体内还发挥着其他作用,包括促进牙齿和骨骼生长以及调整细胞分裂、分化。

30. Vitamin A also plays a role in the regulation of the immune system.

维生素 A 也在调节免疫系统时发挥着作用。

31. Vitamin K is derived from the German word coagulation, which refers to the process of blood clot formation called coagulation.

维生素 K 是从德语单词 coagulation 演化而来,指的是血块形成的过程,称为凝固。

32. It is the main role for vitamin K in humans to be part of the proteins that forms clots to close off a wound.

维生素 K 对人的主要作用是与蛋白质结合,形成伤口愈合的血块。

33. Vitamin K is involved in bone metabolism and calcium absorption because it aids in the re-mineralization of bones, resulting in increased bone density.

维生素 K 参与骨骼新陈代谢和钙的吸收,因为它有助于骨骼的再矿化,从而提高骨质密度。

34. Vitamin K may be important for prevention of heart disease by preventing the hardening of arteries.

通过防止动脉硬化,维生素 K 对于预防心脏疾病可能很重要。

35. There are two main forms of vitamin K: phylloquinone and menaquinone.

维生素 K 有两个主要形态:叶绿醌和甲基萘醌。

36. All cells of the body are surrounded by a membrane that contains fat, cholesterol and various proteins.

所有的细胞都被一层含有脂肪、胆固醇和多种蛋白质的膜包裹着。

37. The wide range of benefits to the body includes a protective effect on cardiovascular disease and cancer.

人们发现维生素 E 对身体有很多益处,包括对于心血管疾病和癌症的预防保护。

38. Vitamin E is a scavenger of oxidation agents in the body.

维生素 E 可以说是人体内氧化性物质的清除专家。

39. Vitamin E collects the free radicals, and provides a defense against tissue damage.

维生素 E 收集自由基,为人体组织提供一道有效的保护屏障。

40. Vitamin D is unique among vitamins because it is produced naturally in the body.

在众多维生素中,维生素 D 之所以独特是因为它可以在体内合成。

41. Sunlight transforms cholesterol which is present in the skin into vitamin D.

阳光照射皮肤时,日光将皮肤里存在的胆固醇转化为维生素 D。

42. Vitamin D is important in bone health because it enhances calcium absorption in the

intestine.

维生素 D 对于骨骼健康很重要,因为它能提高肠道对钙的吸收。

43. Calcium is the building blocks for bones.

钙是骨骼生长的基础材料。

44. Vitamin D also aids the immune system by increasing the body's ability to eliminate microbes which can cause infection.

通过提高人体消灭致感染微生物的能力,维生素 D 还能增强免疫系统的功能。

45. One way to classify the vitamins are as water-soluble or fat-soluble.

划分维生素种类的一种方式就是将其划分为水溶性或脂溶性。

46. The water-soluble vitamins are distributed widely in the body but not stored; therefore, they are not likely to be associated with toxic overdose.

水溶性维生素广泛存在于身体之中,但并不能被储存,因此其不可能因吸收过量而出现中毒的情况。

47. The water-soluble vitamins include vitamin C and the B vitamins (the B-complex).

水溶性维生素也就是维生素 C 和维生素 B(维生素 B 群)。

48. Cobalamin, more commonly referred to as vitamin B12, is useful for maintaining healthy nerve cells and red blood cells.

维生素 B12 又叫作钴胺素,它对保持神经细胞和红细胞的健康具有重要意义。

49. If a person does not have enough vitamin B12 in their diet, they may experience symptoms such as depression, anemia, weakness or fatigue, and neurological problems that include numbness or tingling in the hands or feet.

如果一个人没有摄入足够的维生素 B12,他们就会产生抑郁、贫血、无力或疲倦等症状,同时神经系统也会受到损害,包括手脚麻木和刺痛。

50. Vitamin C functions as an antioxidant in the body to block damages done by free radicals, which could otherwise build up in the body and cause great damage to the body in the long term.

维生素 C 在人体中充当抗氧化剂,这就意味着它能够阻挡自由基带来的损伤。自由基如果在人体聚集,长此以往就会对人造成损害。

Task Two Sample Dialogue

Directions:*In this section, you are going to read several times the following sample dialogue about the relevant topic. Please pay special attention to five C's (culture, context, coherence, cohesion and critique) in the dialogue and get ready for a smooth communication in the coming task.*

A street interview

(*A journalist is interviewing residents about their knowledge of salt.*)

Journalist: Sir, do you know what sodium is?

Resident A: Oh, sodium is something that affects everyone, every day.

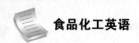

Resident B: Salt is an essential source of sodium in our daily life.

Journalist: Quite right. How much sodium do you usually ingest every day?

Resident A: Oh, not have a clue.

Resident B: Me neither. Normal? I guess.

Journalist: Why can't people live without salt, or, to be specific, sodium?

Resident A: The most important function of salt is to preserve food and slow microbial growth, which extends the shelf-life of perishable items.

Resident B: It enhances the flavor of foods by masking bitterness and acidity and has the ability to make sweets taste sweeter.

Resident C: Sodium is an essential mineral which means it is vital for the human body. It is beneficial to the normal function of cells and organs.

Journalist: Have you ever been worried about the health implications of a high sodium diet?

Resident A: High sodium intake has been linked to high blood pressure, heart disease and stroke.

Resident B: People with kidney disease, diabetes, and middle-aged or older adults should limit their sodium intake.

Journalist: Thank you, Sir. You both seem to be fairly aware of the benefits of sodium in salt, as well as the danger of the high sodium diet. Thanks for your time.

Task Three　Simulation and Reproduction

Directions: *A TV program is having a survey on the following subjects on the street. Any passersby will be interviewed to express their ideas. If you are asked, what's your opinion?*

1) Should we eat more fruit and vegetables?

2) Why should we keep a balanced diet?

Task Four　Discussion and Debate

Directions: *The class will be divided into two groups. Please choose your stand in regard to the following beliefs and try your best to support your opinions with the specialized terms and classical sentences you have learned in the previous parts of this unit.*

The modern technology can analyze the multiple nutrients contained in food scientifically, and it is recommended for people to have a balanced and comprehensive diet. However, in china, such an idea is truly believed by a lot of citizens that one can improve the health of a specific organ by eating food with the same shape, such as having walnut to provide nutrition for brains. Which diet principle are you in favor of?

V. After-class Exercises

1. *Match the English words in Column A with the Chinese meaning in Column B.*

A	B
1）carbohydrate	a. 新陈代谢
2）iodine	b. 视黄醇
3）metabolism	c. 光合作用
4）nourishment	d. 碘
5）photosynthesis	e. 碳水化合物
6）protease	f. 调味
7）retinol	g. 蛋白酶
8）satiety	h. 食品学,营养学
9）season	i. 饱腹感
10）sitology	j. 食物,滋养品

2. *Fill in the following blanks with the words or phrases in the word bank. Some may be chosen more than once. Change the forms if it's necessary.*

polyunsaturated	omega-3	saturated	vitamin D	monounsaturated
complementary	fat-soluble	sustained	carbohydrate	intensity
omega-6	vitamin K	complete	retina	water-soluable
lutein	incomplete	tamper	accessible	

1）Vitamin C and vitamin B are _____ vitamins.

2）_____ can be produced naturally in the body through sunlight.

3）An upper tolerable limit has been defined for each _____ vitamin.

4）Fats can be categorized into _____ fats, _____ fats and _____ fats.

5）To protect from macular degeneration, one should eat more food with _____.

6）_____ can keep one's physical energy much longer than protein.

7）Polyunsaturated fats come in two main classes: _____ and _____.

8）_____ in humans helps to form clots to close off a wound.

9）The categories of proteins include _____, _____ and _____.

10）When two or more _____ proteins are mixed together, they form a complete protein.

3. *Translate the following sentences into English.*

1）健康成人膳食指南推荐成人每天的钠摄入量为 2300 毫克(mg),而高危人群仅为 1500 毫克。

_____

_____

2）地中海居民的饮食中的脂肪大多为单一不饱和脂肪,所以心脏病发病率不高。

_____

_____

3）根据最新报导,糖尿病呈现出年轻化趋势,甚至十几岁的青少年也开始患病。

_____

4）据研究，减少30%的盐摄取量能降低25%的心血管疾病死亡率。

_____

_____

5）越来越多的食品生产商正自愿加入进来，自愿降低其食品中的含盐量。

_____

_____

4. *Your company has decided to release a new food product to the market. Here is the nutrition label of it. As a salesman, after reading the following information, you are going to interpret its contents to the customers. Coherence and comprehensiveness of the information will be highly appreciated.*

Nutrition Facts

Serving Size 1 cup (228g)
Servings Per Container 2

Amount Per Serving

Calories 250   Calories from Fat 110

	% Daily Value*
Total Fat 12g	18%
Saturated Fat 3g	15%
Trans Fat 3g	
Cholesterol 30mg	10%
Sodium 470mg	20%
Total Carbohydrate 31g	10%
Dietary Fiber 0g	0%
Sugars 5g	
Protein 5g	
Vitamin A	4%
Vitamin C	2%
Calcium	20%
Iron	4%

* Percent Daily Values are based on a 2,000 calorie diet. Your Daily Values may be higher or lower depending on your calorie needs.

	Calories:	2,000	2,500
Total Fat	Less than	65g	80g
Sat Fat	Less than	20g	25g
Cholesterol	Less than	300mg	300mg
Sodium	Less than	2,400mg	2,400mg
Total Carbohydrate		300g	375g
Dietary Fiber		25g	30g

_____

_____

_____

_____

_____

_____

_____

_____

_____

_____

_____

_____

_____

_____

VI. Additional Reading

Nestlé Company Profile

Nestlé's origins date back to 1866, when two separate Swiss enterprises were founded that would later form the core of Nestlé. In the succeeding (随后的) decades, the two competing enterprises aggressively expanded their businesses throughout Europe and the United States. In August 1867, Charles (US consul in Switzerland) and George Page, two brothers from Lee County, Illinois, USA, established the Anglo-Swiss Condensed (浓缩的) Milk Company in Cham, Switzerland. Their first British operation was opened at Chippenham, Wiltshire, in 1873. In September 1866 in Vevey, Henri Nestlé developed milk-based baby food, and soon began marketing it. The following year saw Daniel Peter begin seven years of work perfecting his invention, the milk chocolate manufacturing process. Nestlé was the crucial co-operation that Peter needed to solve the problem of removing all the water from the milk added to his chocolate and thus preventing the product from developing mildew (霉). Henri Nestlé retired in 1875 but the company, under new ownership, retained his name as Société Farine Lactée Henri Nestlé.

In 1877, Anglo-Swiss added milk-based baby foods to their products; in the following year, the Nestlé Company added condensed milk to their portfolio (产品组合), which made the firms direct and fierce rivals (竞争对手). In 1879, Nestlé merged with milk chocolate inventor Daniel Peter. In 1904, François-Louis Cailler, Charles Amédée Kohler, Daniel Peter and Henri Nestlé participated in the creation and development of Swiss chocolate, marketing the first chocolate-milk Nestlé.

In 1905, the companies merged (合并) to become the Nestlé and Anglo-Swiss Condensed Milk Company, retaining that name until 1947, when the name "Nestlé Alimentana SA" was taken as a result of the acquisition (并购) of Fabrique de Produits Maggi SA (founded 1884) and its holding company, Alimentana SA, of Kempttal, Switzerland. Maggi was a major manufacturer of soup mixes and related foodstuffs. The company's current name was adopted in 1977. By the early 1900s, the company was operating factories in the United States, the United Kingdom, Germany, and Spain. The First World War created demand for dairy products in the form of government contracts, and, by the end of the war, Nestlé's production had more than doubled.

Nestlé felt the effects of the Second World War immediately. Profits dropped from US $20 million in 1938, to US $6 million in 1939. Factories were established in developing countries, particularly in Latin America. Ironically, the war helped with the introduction of the company's newest product, Nescafé ("Nestlé's Coffee"), which became a staple (主要的) drink of the US military. Nestlé's production and sales rose in the wartime economy.

In February 2014, Nestlé sold its PowerBar sports nutrition business to Post Holdings, Inc.. Later, in November 2014, Nestlé announced that it was exploring strategic options for its frozen food subsidiary (附属机构), Davigel.

In recent years, Nestlé Health Science has made several acquisitions. It acquired Vitaflo, which makes clinical nutritional products for people with genetic disorders; CM&D Pharma Ltd., a company that specialises in the development of products for patients with chronic conditions like kidney disease; and Prometheus Laboratories, a firm specializing in treatments for gastrointestinal (胃肠的) diseases and cancer. It also holds a minority stake in Vital Foods, a New Zealand-based company that develops kiwi fruit-based solutions for gastrointestinal conditions.

In December 2014, Nestlé announced that it was opening 10 skin care research centers worldwide, deepening its investment in a faster-growing market for healthcare products. That year, Nestlé spend about $350 million on dermatology research and development. The first of the research hubs (中心), Nestlé Skin Health Investigation, Education and Longevity Development (SHIELD) centers, will open mid 2015 in New York, followed by Hong Kong and São Paulo, and later others in North America, Asia and Europe. The initiative is being launched in partnership with the Global Coalition on Aging (GCOA), a consortium (财团) that includes companies such as Intel and Bank of America.

Logo

The original Nestlé trademark was based on his family's coat of arms, which featured a single bird sitting on a nest. This was a reference to the family name, which means "nest" in German. Henri Nestlé adapted the coat of arms by adding three young birds being fed by a mother, to create a visual link between his name and his company's infant cereal products. He began using the image as a trademark in 1868. Today, the familiar bird's nest logo continues to be used on Nestlé products worldwide, in a modified(改良的) form.

Objective

Nestlé's objective is to consolidate (巩固) and strengthen its leading position at the cutting edge of innovation in the food area to meet the needs and desires of customers around the world, for pleasure, convenience, health and well-being.

Core Values

"The Nestlé global vision is to be the leading Health, Wellness, and Nutrition Company in the world."

Mission Statements

"...positively influence the social environment in which we operate as responsible corporate citizens, with due regard for those environmental standards and societal aspirations which improve quality of life." — Henri Nestlé, 1857

Vision Statement

To be a leading, competitive Nutrition, Health and Wellness Company delivering improved shareholder value by being a preferred corporate citizen, preferred employer, preferred supplier selling preferred products.

Nestlé's Commitment

Quality and safety for our consumers is Nestlé's top priority. This applies to our entire portfolio, from foods and beverages (饮料) to all our systems and services. Quality assurance and product safety is one of Nestlé's 10 Corporate Business Principles which form the foundation of all we do.

Product

Nestlé has over 8 000 brands with a wide range of products across a number of markets, including coffee, bottled water, milkshakes and other beverages, breakfast cereals, infant foods, performance and healthcare nutrition, seasonings, soups and sauces, frozen and refrigerated foods, and pet food.

Main Brand Crises

1) Maggi Noodles

In May 2015, Food Safety Regulators from the Uttar Pradesh, India found that samples of Nestlé's leading noodles Maggi had up to 17 times beyond permissible safe limits of lead in addition to monosodium glutamate (味精).

On 3 June 2015, New Delhi Government banned the sale of Maggi in New Delhi stores for 15 days because it found lead and monosodium glutamate in the eatable beyond permissible limit. The Gujarat FDA on 4 June 2015 banned the noodles for 30 days after 27 out of 39 samples were detected with objectionable (令人反感的) levels of metallic (金属的) lead, among other things. Some of India's biggest retailers like Future Group, Big Bazaar, Easyday and Nilgiris have imposed a nationwide ban on Maggi. Thereafter multiple state authorities in India found unacceptable amount of lead and it has been banned in more than 5 other states in India.

2) Milk Products and Baby Food

In late September 2008, the Hong Kong government found melamine (三聚氰胺) in a Chinese-made Nestlé milk product. Six infants died from kidney damage, and a further 860 babies were hospitalized (送……住院). The Dairy Farm milk was made by Nestlé's division in the Chinese coastal city Qingdao. Nestlé affirmed that all its products were safe and were not made from milk adulterated (掺假) with melamine. On 2 October 2008, "the Taiwan Health Ministry" announced that six types of milk powders produced in the Chinese mainland by Nestlé contained low-level traces of melamine, and were "removed from the shelves". In another incident, weevils (象鼻虫) and fungus (真菌) were found in Cerelac baby food.

Nestlé has implemented initiatives to prevent contamination (污染) and utilizes what it calls a "factory and farmers" model that eliminates (排除) the middleman. Farmers bring milk directly to a network of Nestlé-owned collection centers, where a computerized system samples, tests, and tags each batch of milk. To reduce further the risk of contamination at the source, the company provides farmers with continuous training and assistance in cow selection, feed quality, storage, and other areas.

3) Cookie Dough

In June 2009, an outbreak of E. coli was linked to Nestlé's refrigerated cookie dough (生面团) originating in a plant in Danville, Virginia. In the US, it caused sickness in more than 50 people in 30 states, half of whom required hospitalization. Following the outbreak, Nestlé voluntarily recalled 30 000 cases of the cookie dough. The cause was determined to be contaminated flour obtained from a raw material supplier. When operations resumed, the flour used was heat-treated to kill bacteria.

4) Bottled Water

At the second World Water Forum, Nestlé and other corporations persuaded the World Water Council to change its statement so as to reduce access to drinking water from a "right" to a "need". Nestlé chairman and former CEO Peter Brabeck-Letmathe stated that "access to water should not be a public right". Nestlé continues to take control of aquifers (含水层) and bottle their water for profit. Peter Brabeck-Letmathe has later changed his statement.

The Nescafé Plan

In 2010, Nestlé launched the Nescafé Plan, an initiative to increase sustainable coffee production and make sustainable coffee farming more accessible to farmers. The plan aims to increase the company's supply of coffee beans without clearing rainforests, as well as using less water and fewer agrochemicals. According to Nestlé, Nescafé will invest 350 million Swissfrancs (about $336 million) over the next ten years to expand the company's agricultural research and training capacity to help benefit many of the 25 million people who make their living growing and trading coffee. The Rainforest Alliance and the other NGOs in the Sustainable Agriculture Network will support Nestlé in meeting the objectives of the plan.

Main Focus of Nestlé

1) People First

Employees, people and products are more important at Nestlé than systems. Systems and methods, while necessary and valuable in running a complex organization, should remain managerial (管理上的) and operational aids but should not become ends in themselves. It is a question of priorities. A strong orientation toward human beings, employees and executives is a decisive, if not the decisive, component of long-term success. Nestlé is very much keen about talent management and the HR department works to hire and retain the best industry personnel.

2) Quality Products

Their focus is on products. The ultimate justification for a company is its ability to offer products that are appealing because of their quality, convenience, variety and price—products that can stand their ground even in the face of fierce competition.

3) Long-term View

Nestlé makes a clear distinction between strategy and tactics. It gives priority to the long-range view. Long-term thinking defuses many of the conflicts and contentions (竞争) among groups—this applies to employment conditions and relations with employees as well as to the conflicts and opposing interests of the trade and the industry. Of course, the ability to focus on long-term considerations is only possible if the company is successful in the struggle for short-term survival. This is why Nestlé strives to maintain a satisfactory level of profits every year.

4) Diversification

Nestlé does not want to become either a conglomerate (联合大企业) or a portfolio manager. Nestlé wants to operate only those businesses about which it has some special knowledge and expertise. Nestlé is a global company, not a conglomerate hodgepodge (大杂烩). It regard acquisitions and efforts at diversification as logical ways to supplement (增补) the business, but only in the context of a carefully considered corporate marketing policy.

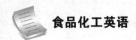

5) Research and Development

Nestlé is probably unique in the food industry in having an integrated research and development program that engages in applied and basic research in the fields of human physiology, health, nutrition and raw materials. The research and development program makes it possible to create new types of products that cannot even be imagined today, especially in the critical area where preventive medicine and food products overlap.

For Nestlé, this is particularly important in packaging. Concern for the effects of packaging on the environment is forcing it to look for new solutions and to consider their interaction with the biological product—food.

Nestlé is one of the world's leading food companies and intends to remain so. Its commitment to high quality market research ensures that it remains fully aware of changes in consumer behavior and consumer tastes. Its excellent product research and development network ensures that it is well placed to meet the challenge of changes in consumer expectations. The company's wellness strategy is carefully geared to(调整使其适合……) delivering to customers what they now clearly want in relation to the foods they eat, a high nutritional value and a positive contribution to their general wellness.

(*For more information*: *https*: //*www.Nestlé.com/aboutus/history/logo-evolution*; *https*: //*bohatala. com/nestle-company-profile/*; *https*: //*www. Nestlé. com/aboutus/history/logo-evolution*)

1. *Read the passage quickly by using the skills of skimming and scanning. And choose the best answer to the following questions.*

   1) In which country is the headquarters of Nestlé Company located?

      A. America.   B. Switzerland.

      C. Britain.   D. Germany.

   2) Which statement about Henri Nestlé is not true?

      A. He developed milk-based baby food, and soon began marketing it.

      B. He spent seven years of work perfecting his invention.

      C. He retired in 1875.

      D. After his retirement, the company, under new ownership, retained his name.

   3) By the early 1900s, the company was operating factories in the following countries except _____.

      A. the United States   B. Spain

      C. the United Kingdom   D. China

   4) Which of the following statements about Nestlé's is true?

      A. The Second World War had no impact on Nestlé.

      B. Profits increased from US $6 million in 1938, to US $20 million in 1939.

      C. By the end of the war, Nestlé's production had more than doubled.

      D. The war prohibited the introduction of the company's newest product, Nescafé.

5) In November 2014, Nestlé announced that it was exploring strategic options for Davigel. Davigel is _____.

A. a food company that is owned by a larger company

B. a sort of nutrient

C. a recipe

D. a marketing strategy

6) Which one is not acquired by Nestlé Health Science in recent years?

A. Vitaflo.　　　　　　　　　　B. CM&D Pharma Ltd.

C. Prometheus.　　　　　　　　　D. Mars Company.

7) The original Nestlé trademark was based on _____.

A. family's coat of arms　　　　　B. the founder's interest

C. the founder's major　　　　　　D. an employee's advice

8) What's Nestlé's top priority according to its business commitment?

A. Quality and safety for our consumers.　B. Profits.

C. Commercial competiveness.　　　D. Diversity in products.

9) Which of the following product was banned by New Delhi Government?

A. Cookie dough.　　　　　　　　B. Breakfast cereals.

C. Maggi noodles.　　　　　　　　D. Bottled water.

2. *Match the time in Column A with the corresponding development achievement in Column B.*

A	B
1866	1) The Anglo-Swiss Condensed Milk Company was established.
1867	2) The company's current name was adopted.
1877	3) Anglo-Swiss added milk-based baby foods to their products.
1879	4) Companies merged to become the Nestlé and Anglo-Swiss Condensed Milk Company.
1905	5) Nestlé merged with milk chocolate inventor Daniel Peter.
1977	6) Two separate Swiss enterprises, the core of Nestlé were founded.

3. *In this part, you are going to make an oral presentation on either of the following topics.*

1) A summary of the basic information of Nestlé.

2) Suggestions for Nestlé to deal with the food trust crises.

习题答案

Unit Three　Chemical Reactions

I. Pre-class Activity

Directions：*Please read the general introduction about Sir William Ramsay and tell something more about the great scientist to your classmates.*

Sir William Ramsay

Sir William Ramsay (2 October 1852－23 July 1916) was a Scottish chemist who discovered the noble gases(惰性气体) and received the Nobel Prize in Chemistry in 1904 "in recognition of his services in the discovery of the inert gaseous elements(惰性气体元素)in air". After the two men identified argon(氩), Ramsay investigated other atmospheric gases. His work in isolating argon,helium(氦),neon(氖),krypton(氪) and xenon(氙) led to the development of a new section of the periodic table.

In 1887,he succeeded(接任)Alexander Williamson as the chair of Chemistry at University College London (UCL). It was here at UCL that his most celebrated(著名的) discoveries were made. As early as 1885－1890,he published several notable papers on the oxides of nitrogen(氮的氧化物),developing the skills that he needed for his subsequent (随后的) work. On the evening of 19 April 1894,Ramsay attended a lecture given by Lord Rayleigh. Rayleigh had noticed a discrepancy(矛盾,不符合) between the density (浓度) of nitrogen made by chemical synthesis(合成) and nitrogen isolated(分离) from the air by removal of the other known components. After a short conversation,he and Ramsay decided to investigate this. In August,Ramsay told Rayleigh he had isolated a new,heavy component of air,which did not appear to have any chemical reactivity(化学反应性). He named this inert gas "argon",from the Greek word meaning "lazy". In the following years,working with Morris Travers,he discovered neon,krypton,and xenon. He also isolated helium,which had only been observed in the spectrum(光谱) of the sun,and had not previously found on earth. In 1910,he isolated and

characterized radon(氡)

He died in High Wycombe, Buckinghamshire, on 23 July 1916 from nasal(鼻腔的) cancer at the age of 63 and was buried in Hazlemere Parish Church.

II. Specialized Terms

Directions: *Please memorize the following specialized terms before the class so that you will be able to better cope with the coming tasks.*

abrade v.磨损

absorbent n.吸收剂

acidic adj.酸性的

additive n.添加剂

adhesive n.胶黏剂

agent n.药剂

alkaline adj.碱性的

allotrope n.同素异形体

alloy n.合金

alumina n.矾土

amide n.酰胺

amine n.胺

ammonia n.氨水

aniline n.苯胺

antioxidant n.抗氧化剂

aqua n.水

bleach n.漂白剂

brominate v.使溴化

butane n.丁烷

catalyst n.催化剂

caustic adj.腐蚀性的

chitosan n.壳聚糖

chlorinated adj.用氯消毒的

chloroform n.氯仿

choline n.胆碱

chromia n.氧化铬

colorant n.着色剂

composite n.复合物

condense v.凝结

contaminant n.污染物

corrosive adj.腐蚀性的

crystal n.晶体

decomposition n.分解

degradation n.降解

deionization n.消电离

density n.密度

deposition n.沉淀

desiccant n.干燥剂

ethanolamine n.乙醇胺

diffuse v.扩散

dilute v.稀释

dioxane n.二氧六环

dissolve v.溶解

elasticity n.弹性

electrode n.电极

electrolysis n.电解

electrolyte n.电解液

epoxy n.环氧树脂

ethane n.乙烷

ethanol n.乙醇

ether n.乙醚

ethylene n.乙烯

extraction n.提取

ferment v.发酵

filter n.过滤器

flammable adj.易燃的

fluorescence n.荧光

flux n.熔剂,助熔剂

formaldehyde n.甲醛

formulation n.配方

gel n.凝胶

glycol n.乙二醇

graphene n.石墨烯

graphite n.石墨

greaseless adj.没有油脂的

hexagonal adj.六角形的

hormone n.激素

hydrocarbon n.烃

hydrochloric acid 盐酸

hydrofluoric n.氟化氢

lattice n.晶格

loop n.环路

lubricant n.润滑剂

matrix n.基体

nitrobenzene n.硝基苯

nitrosamine n.亚硝胺

oxidation n.氧化

phosphate n.磷酸盐

phosphoric adj.磷的

phthalate n.酞酸酯

polycyclic adj.多环的

polydisperse adj.多分散的

polyester n.聚酯

polyethylene n.聚乙烯

polymer n.聚合物

polyolefin n.聚烯烃

polystyrene n.聚苯乙烯

polyurethane n.聚氨酯

potash n.碳酸钾

quantify v.量化

radioactive adj.有辐射的

reagent n.试剂

refine v.提炼

resin n.树脂

retardant n.阻滞剂

sensibility n.感受性,敏感性

silicate n.硅酸盐

smelt v.冶炼

sodium laureth sulfate 十二烷基硫酸钠

solvent n.溶剂

III. Watching and Listening

Task One　Balancing Equations

New Words

balance v.(使)平衡

equation n.方程式

mole n.分子(molecule 的缩写)

concept n.概念

fairly adv.相当地

straightforward adj.直接的

diatomic adj.二原子的,二氢氧基的,二价的

oxide n.氧化物

reactant n.反应物

multiply v.乘以

yield v.产生,生产

tempt v.诱使

tweak v.稍微调整

ratio n.比率

methane n.甲烷

parenthesis n.括号

engineer v.安排,处理调整

magenta adj.紫红色或洋红色的

视频链接及文本

Exercises

1. *Watch the video for the first time and choose the best answers to the following questions.*

1) What will it end up with if there are some aluminum and some oxygen gas in a chemical reaction?

    A. Water.   B. Oxygen.

    C. Aluminum oxide.   D. Carbon dioxide.

2) Why is it difficult for the students to balance the equation?

    A. Because it is an abstract concept.

    B. Because it involves complicated calculation.

    C. Because it requires some arts in balancing the equations.

    D. Because it is a new subject for the students.

3) Which of the following is correct about balancing the equations?

    A. The number of the same atoms at both ends should be the same.

    B. The number of the same molecules at both ends should be the same.

    C. There must be the same number of substances at both ends of the equation.

    D. The state of substances at both ends should agree with each other.

4) All the following numbers can be added to a molecule except _____ in the process of balancing chemical equations.

    A. 2   B. 3

    C. 3.5   D. 6

5) In the following chemical reaction: $C_2H_4(g)+3O_2(g)=2CO_2(g)+2H_2O(l)$, the letter "g" in the parenthesis signifies _____.

    A. the weight of the substance

    B. the temperature of the substance

    C. the state of the substance

    D. the property of the substance

2. *Watch the video again and decide whether the following statements are true or false.*

1) In the following chemical reaction: $C_2H_4+3O_2\rightarrow 2CO_2+2H_2O$, the reactants are CO_2 and H_2O. (　)

2) The equation: $Al+O_2=Al_2O_3$ is imbalanced. (　)

3) There is no mistake in the equation: $H_2(g)+O_2(g)=H_2O(g)$. (　)

4) In balancing an equation, the molecule H_2O should be taken as a whole. (　)

5) If the number of a certain atom is 1.5 in the process of balancing, we can multiply it by 2 and then get 3. (　)

3. *Watch the video for the third time and fill in the following blanks.*

    Like the concept of the mole, balancing equations is one of those ideas that you learn in first-year 1) _____ class. It tends to give a lot of students a hard time, even though it is a fairly 2) _____ concept. I think what makes it 3) _____ is that there's a bit of an 4) _____ to it. So before we talk about balancing 5) _____ equations, what is a chemical equation?

Well here are some examples right here, and I have some more in the rest of this video. But it 6) _____ just describes a chemical reaction. You've got some 7) _____. You have some oxygen gas or a diatomic oxygen 8) _____. And then you end up with aluminum oxide. And you would say, ok, fine, that's an equation. It looks nice. I have my 9) _____, or the things that react. These are the reactants. And then I have the 10) _____ of this reaction…

4. *Share your opinions with your partners on the following topics for discussion.*

1) Please describe to the class the composition of a chemical equation. An example will be appreciated in making such a presentation.

2) Do you think it is an easy job to balance the chemical equations? Do you have some unique secrets in accomplishing such an abstract task? Please share them with your classmates.

Task Two　Atom Bonds

New Words

individual adj.个人的,单个的
bond n.纽带;[化]键
stick v.粘贴
essentially adv.本质上,根本上
collection n.收集,采集
bunch n.束,串,捆
stable adj.稳定的
configuration n.[化]（分子中原子的）组态,排列
shell n.壳
envious adj.羡慕的,嫉妒的
alkali n.碱
offload v.卸下,卸货
halogen n.卤素
electronegativity n.电负性

flavor v.给……调味
valence n.化合价
notation n.记号,标记法
Coulomb force 库仑力
ionic bond 离子键
season v.调味
ion n.离子
cation n.阳离子
anion n.阴离子
necessarily adv.必须地
notion n.概念,观念
probability n.概率
distribution n.分布
covalent adj.共价的
covalent bond 共价键

视频链接及文本

Exercises

1. *Watch the video for the first time and choose the best answers to the following questions.*

1) If a certain number of atoms stick together in a certain way, a _____ is produced.

　　A. molecule　　　　　　　　　　B. substance

C. nucleus   D. protein

2) The reason why an atom wants to give an electron in the process of chemical reaction is
that _____.

A. it has too many electrons around its nucleus

B. it wants to carry a positive charge

C. it wants to achieve a stable configuration

D. it aims to develop into another type of atom

3) According to the video clip, other atoms are envious of the noble gases because _____.

A. they have no electrons in the outer shell

B. they have eight electrons in the outer shell

C. they are widely used in the modern industry

D. they are very lazy and don't react easily

4) If sodium is put together with _____, we will have the table salt.

A. chlorine   B. carbon

C. calcium   D. charcoal

5) When an atom gives an electron, it will carry _____ charge; and the atom will carry
_____ charge when it takes some electrons.

A. positive, positive   B. positive, negative

C. negative, negative   D. negative, positive

2. *Watch the video again and decide whether the following statements are true or false.*

1) When a chlorine atom takes an electron, it will develop into an argon atom.(   )

2) In NaCl, Na^+ is a cation while Cl^- is an anion.(   )

3) In an ionic bond, the two atoms involved will share a certain number of electrons.(   )

4) In a chemical molecule O_2, the two oxygen atoms will be stuck together through an
ionic bond.(   )

5) When two atoms are connected through a covalent bond, they will share a certain
number of electrons in the outer shell.(   )

3. *Watch the video for the third time and fill in the following blanks of the table.*

Chemical Symbols	Number of Protons	Number of Electrons involved	Type of Bond	Positive/Negative Charge
Na^+ in NaCl				
Cl^- in NaCl				
Al^{3+} in Al_2O_3				
O in O_2				
O^{2-} in H_2O				

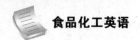

4. *Share your opinions with your partners on the following topics for discussion.*

1) How many different ways have been touched in the video about linking two atoms together? Please summarize the main idea in your own words. Be sure to use the specialized terms.

2) It is said that there is a third bond between atoms—metallic bond(金属键). Please log onto the internet and search relevant information about it. Some of you will be asked to give a presentation. Specific examples will be highly appreciated.

IV. Talking

Task One   Classical Sentences

Directions：*In this section，some popular sentences are supplied for you to read and to memorize. Then，you are required to simulate and produce your own sentences with reference to the structure.*

General Sentences

1. —What was the weather like yesterday?

—Yesterday it rained all day.

——昨天天气怎样？

——昨天下了一天的雨。

2. —What will the weather be like tomorrow?

—It's going to snow tomorrow.

——明天天气怎样？

——明天有雪。

3. Do you think it's going to rain tomorrow?

你觉得明天会下雨吗？

4. I don't know whether it will rain or not.

我不知道明天会不会下雨。

5. It'll probably clear up this afternoon.

今天下午可能会放晴。

6. The days are getting hotter.

天气在变暖。

7. What's the temperature today?

今天多少度？

8. It's about twenty degrees centigrade this afternoon.

今天下午大约二十摄氏度。

9. There's a cool breeze this evening.

今晚有股冷风。

10. Personally, I prefer winter weather.

就我个人而言, 我比较喜欢冬天的天气。

11. What time are you going to get up tomorrow morning?

明天早上你打算几点起床？

12. I'll probably wake up early and get up at 6:30.

我可能会醒得比较早, 大概六点半起床。

13. —What will you do then?

—After I get dressed, I'll have breakfast.

——接着你做些什么呢？

——穿上衣服后我就去吃早饭。

14. What will you have for breakfast tomorrow morning?

明天早餐你吃什么呢？

15. I'll probably have eggs and toast for breakfast.

早餐我可能会吃土司面包和鸡蛋。

16. After breakfast, I'll get ready to go to work.

吃完早餐后我会准备一下去上班。

17. I get out of bed about 7 o'clock every morning.

每天早上我 7:00 起床。

18. After getting up, I go into the bathroom and take a shower.

起床后我会去浴室冲个澡。

19. Then, I brush my teeth and comb my hair.

然后我刷牙梳头。

20. After brushing my teeth, I put on my clothes.

我刷完牙后穿衣服。

21. After that, I go downstairs to the kitchen to have breakfast.

然后, 我下楼到厨房去吃早饭。

22. I'll leave the house at 8:00 and get to the office at 8:30.

我 8 点离开家, 8 点半到办公室。

23. I'll probably go out for lunch at about 12:30.

我 12:30 左右去吃午饭。

24. I'll finish working at 5:30 and get home by 6 o'clock.

我 5:30 下班, 6:00 到家。

25. I'm always tired when I come home from work.

下班以后我总是很累。

26. Are you going to have dinner at home tomorrow night?

明天晚上你在家吃晚饭吗？

27. Do you think you'll go to the movies tomorrow night?

明天晚上你会去看电影吗？

28. I'll probably stay home and watch television.
我可能呆在家看电视。

29. I'm not accustomed to going out after dark.
我不习惯晚上出去。

30. When I get sleepy, I'll probably get ready for bed.
当我感到困的时候,我就准备上床睡觉。

31. Do you think you'll be able to go to sleep right away?
你觉得你现在就能去睡觉吗?

32. What do you plan to do tomorrow?
明天你打算干什么?

33. I doubt that I'll do anything tomorrow.
明天恐怕我什么也不做。

34. I imagine I'll do some work instead of going to the movies.
我想做点事,不想去看电影。

35. Will it be convenient for you to explain your plans to him?
你把你的计划跟他讲一下,方便吗?

36. What's your brother planning to do tomorrow?
你哥哥明天计划干什么?

37. It's difficult to make a decision without knowing all the facts.
不知道全部事实而去做决定是很难的。

38. I'm hoping to spend a few days in the mountains.
我想在山上呆几天。

39. Would you consider going north this summer?
你今年夏天想去北方吗?

40. If there's a chance you'll go, I'd like to go with you.
如果你有机会去的话,我想和你一起去。

41. After you think it over, please let me know what you decide.
在你仔细考虑之后,请告诉我你的决定。

42. Are you going to go anywhere this year?
今年你打算去哪儿?

43. If I have enough money, I'm going to take a trip abroad.
要是我有足够的钱,我打算出国旅行。

44. How are you going? Are you going by boat?
你打算怎样去? 是不是乘船去?

45. It's faster to go by plane than by boat.
坐飞机比坐船快。

46. What's the quickest way to get there?
到那儿去最快的交通方式是什么?

47. Altogether it will take ten days to make the trip.

这次旅行总共要花十天时间。

48. It was a six-hour flight/journey/travel/voyage.

这是一次六小时的旅程。

49. I'm leaving tomorrow, but I haven't packed my suitcases yet.

我明天出发,可是我的箱子到现在还没有整理好。

50. I hope you have a good time on your trip.

祝你旅途愉快。

Specialized Sentences

1. Everything we've been dealing with so far has just been with the individual atoms.

迄今为止我们一直讲的都是单个的原子。

2. It tends to give a lot of students a hard time, even though it is a fairly straightforward concept.

尽管这是个相当直接的概念,但可能让许多学生感到头疼。

3. So what you do is you just multiply this so that you end up with whole numbers.

因此,你需要做的是乘以这个数字后,得到的都是整数。

4. You can't change the relative ratios of the aluminum and the oxygen within the aluminum oxide molecule.

你无法改变氧化铝分子中铝原子和氧原子的相对比率。

5. We're trying to engineer how many carbons we have on both sides of this equation.

我们正微调方程式两端碳原子的数量。

6. When you go step by step, it should proceed fairly smoothly.

当我们一步一步做下去,那就会进展得相当顺利了。

7. Remember, I wanted to do the oxygens last, because I can just set this to whatever I want it to be without messing up any of the other atoms.

记住,我想最后处理氧原子,因为我想怎样处理它都不会把其他原子搞乱。

8. And if we started talking about organic chemistry, you'd have a bunch of atoms, a lot of carbons and hydrogens and other things, fitting together and they'd be forming proteins.

而且如果我们开始讨论有机化学,你就会有一些原子,很多碳原子和氢原子等等。它们组合在一起就会构成蛋白质。

9. An atom wants to give an electron because it's trying to get into a stable configuration in its outer shell.

一个原子想释放电子,因为它想在最外电子层实现稳定状态。

10. They're all envious of the noble gases, because the noble gases have eight electrons in their outer shell.

它们都很嫉妒惰性气体,因为惰性气体原子的最外层有八个电子。

11. But for the sake of our purposes, let's say we just have these two atoms.

但是,为了我们的目的,我们设想就只有这两个原子。

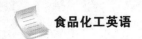

12. It has a negative charge now because it got that electron from sodium.

它现在带有负电,因为它得到了钠的那个电子。

13. What happens in the situation where they're both not as extreme in their views in whether or not they want to give or take electrons?

如果它们双方关于它们是否想得到或释放电子的观点都不那么极端,会出现什么状况呢?

14. When the electron is donated from sodium to chlorine, an ion is formed.

当电子从钠原子转移到氯原子时,就形成了离子。

15. The sodium becomes a cation, because it's positive.

钠成为了阳离子,因为它带正电荷。

16. The chlorine becomes an anion, because it's negative.

氯成为了阴离子,因为它带负电荷。

17. And then cation and anion stuck to each other, so this is an ionic bond.

然后阴离子和阳离子联系在了一起,因此这就是离子键。

18. Now what happens if we have two elements that aren't that different in how much they want electrons?

现在,如果两个元素想得到电子的程度没有那么大的差异,会出现什么情况呢?

19. The best example of that is that we had two of the same element.

关于这一点,最好的例子就是我们有两个相同的元素。

20. Now both of these oxygen atoms would love to have eight electrons.

现在,这两个氧原子都想有八个电子。

21. They could start pretending like they're a noble gas.

它们可以开始假装它们都是惰性气体了。

22. Why don't we share some electrons? And then we can both pretend that we have eight electrons.

我们为什么不共享一些电子呢?这样我们就都可以假装我们有八个电子了。

23. Oxygen doesn't necessarily have to change colors.

氧原子不一定非要改变颜色的。

24. I'm just going to draw this electron over on this side just so you recognize that this is different from other electrons.

我把这个电子画在这边,以便你们能意识到这个电子和其他的电子不一样。

25. And we could do it by drawing a line here.

我们可以在这里画一条线,以达到这个目的。

26. But it can kind of pretend that it has this electron and that electron.

但是它可以假装它拥有了这个电子和那个电子。

27. And the reason why I'm not talking about the noble gases here is that these don't form covalent bonds.

而且我在这里没有讨论惰性气体的原因是它们没有形成共价键。

28. Hydrogen feels good because the one s-shell is completely filled.

氢原子感觉很棒,因为它唯一的 s 层已经完全填满了电子。

29. So in this situation, the electrons are going to spend more time around oxygen than they will around hydrogen.

因此,在这种情况下,这些电子在氧原子周围的时间将比在氢原子周围的时间多。

30. But it's polar, because the electrons are getting pulled to spend most of their time at one side of the atom.

但它是有极性的,因为电子对被拉着偏向某个原子的一端。

31. Now the last bond we can talk about, and I've touched on this a little bit, is the metallic bond.

现在我们能讨论而且我已经提到一点的最后一类键就是金属键。

32. I was in a metallic bond in high school, but anyway, that's a subject for another video.

我上高中时研究过金属键,但是这是另外一集视频的内容了。

33. What makes something metallic or have metallic characteristics is that they have a bunch of electrons in their outer orbital that they're very giving.

真正使它成为金属或是具有金属特性的是它们的最外层轨道有许多活泼的电子。

34. Because their electrons are all on the sea, they've kind of gotten this positive charge.

由于它们的电子都在这个海洋里,因此它们都带正电荷。

35. And this is essentially what allows, well definitely, metals to be conductive, because you have this pool of electrons that are very easy to move around.

显而易见,这就是导致金属能导电的本质,因为形成一个电子很容易自由移动的电子海。

36. If you are to try to bend a bar of salt, the bond will just be broken.

如果你想掰碎一块盐,化学键也就断裂了。

37. This is super useful, because in the rest of chemistry, everything we do will essentially involve some combination of these bonds.

这点超级有用,因为在学习其他化学知识的过程中,所有我们要做的事本质上都牵扯到这些化学键的组合。

38. And we'll start talking about what these bonds mean in terms of the temperature at which they boil, or the properties of the molecules themselves.

今后我们要开始讨论这些化学键在沸点以及分子特性方面的意义。

39. That's NH_3 and it's a gas, that's why the g is in parentheses.

那是 NH_3,是气体,这就是为什么括号内有一个 g。

40. That reaction produces some nitrogen monoxide.

这个反应产生了一些一氧化氮。

41. So the important thing first is to just make sure we have a balanced equation before we even start anything.

因此,首先的一个重要的事情是,我们开始任何事情之前,首先要确保方程式是配

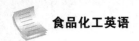

平的。

42. But, sometimes when they talk about oxygen, you have to make sure whether it is molecular or atomic oxygen.

但是,有时,当他们谈到氧气时,你必须弄清楚到底是氧原子还是氧分子。

43. The atomic mass number of oxygen is 16 and I can confirm that by looking at the periodic table down here.

氧的原子质量数是 16,我可以通过查阅下面的元素周期表加以确认。

44. So what's the molecular mass of the diatomic molecule O_2? Well it has 2 oxygen, so it's going to be 2 times 16, which equals 32 atomic mass units.

那么两个氧原子的氧分子质量是多少? 呃,有两个氧原子,所以是 2 乘以 16 等于 32 个单位的相对原子质量。

45. So oxygen is going to be the limiting reagent in this reaction.

因此,氧气成为这个反应里的限制量。

46. So given that we have 1 mole of oxygen, how many moles of ammonia can it react with that?

因此,如果我们有 1 摩尔的氧气,它能与多少摩尔的氨气发生反应呢?

47. I think the hard part is just the conversions between moles and grams.

我认为最难的部分就是摩尔和克之间的转换。

48. Neon really wants to keep that 10th electron, because it fills out the second shell.

氖真的想留住第 10 个电子,因为它可以填满第二个电子层了。

49. You might not know yet, if you haven't seen a covalent bond.

如果你没有见过共价键,你或许还不知道。

50. Now if we wanted to figure out the trend of electronegativity on the periodic table, what do you think is going to happen?

现在,如果我们想找出元素周期表中元素电负性的变化规律,你觉得这个规律会是怎样的?

Task Two　Sample Dialogue

Directions: *In this section, you are going to read several times the following sample dialogue about the relevant topic. Please pay special attention to five C's (culture, context, coherence, cohesion and critique) in the dialogue and get ready for a smooth communication in the coming task.*

In a chemical laboratory

(A teacher and his students are talking about the rules in the chemical laboratory.)

Teacher：　Good morning, boys and girls. Welcome to my class. Today is the first time for you to come into a chemical laboratory. Do you have any idea about the rules in this lab?

Student A: We must keep quiet and we should not move the desks or chairs randomly.

Student B: We should protect the utensils and equipment in this room.

Student C: Oh, pets are not allowed to come in.

Teacher: Besides the ordinary rules, we should bear in mind that chemical labs are unique in some ways. Who'd like to be the first speaker?

Student A: It is filled with various glass utensils, including test tubes and containers of other forms. They are very easy to break. Hence, we should watch out that we may be hurt by the broken glass.

Student B: Also, different chemical substances are stored in the lab. If they are mixed with another gas or liquid, explosion may happen, or some poisonous gas may be produced. Anyhow, it will be very dangerous if we are not careful enough.

Student C: Yes, we should keep the windows open during our experiments, so as to let in fresh air. Also, we must wash our hands before and after doing experiments, because the dirty hands will influence the result of the chemical experiments.

Student D: What's even worse, the chemical substances will make your hands dirty or wounded if you don't keep your hands clean. Please wear a pair of gloves before the experiments.

Teacher: Is there anything else that we should be cautious of?

Student A: No food or drinks are allowed to be taken into the lab. It may make the room dirty and it will be harmful to your health if you go on eating the polluted food.

Student B: We are prohibited from bringing lighters or matches into the lab. If we are not careful enough, a huge fire may break out.

Student C: We must wear masks over our faces when we are doing experiments in this chemical lab. In this way, we can prevent the poisonous air from entering our noses.

Student D: Guys, we should put on that long white uniform, which is a typical symbol of a doctor in the hospital. If we carelessly spill some dangerous substances onto our fashionable jeans, a hole will be made.

Teacher: To my amazement, my dear students have rich knowledge about such a laboratory. I will summarize your viewpoints in the following ways. All in all, we should keep three key words in mind: personal safety, lab environment, and experiment results. As long as you make the above three points work well, I am sure you will enjoy doing various chemical experiments in this lab. Thank you all.

Task Three　Simulation and Reproduction

Directions: *The class will be divided into three major groups, each of which will be assigned a topic. In each group, some students may be the teacher, while others may be students. In the process of discussion, please observe the principles of cooperation, politeness and choice of words. One of the*

groups will be chosen to demonstrate the discussion to the class.

1）The importance of doing experiments in learning chemistry.

2）How to do group work well in doing chemical experiments.

3）The application of noble gas in our daily life.

Task Four　Discussion and Debate

Directions：*The class will be divided into two groups. Please choose your stand in regard to the following controversy and support your opinions with scientific evidences. Please refer to the specialized terms and classical sentences in the previous parts of this unit.*

In the modern chemistry, some chemists believe that chemical reactions depend on the surrounding factors, such as temperature, humidity and pressure. On the other hand, another group of people hold the opinion that, if you mix some substances together, they will naturally produce something new, regardless of the above three factors. Which party do you agree with? Why?

V. After-class Exercises

1. *Give the Chinese meaning to the following specialized terms, as well as the verbs from which the words derive.*

	Chinese Meaning	Verbs	Verbs' Chinese Meaning
absorbent	吸收剂	absorb	吸收
additive			
adhesive			
oxidizer			
colorant			
catalyst			
desiccant			
solvent			
lubricant			
retardant			

2. *Fill in the following blanks with the words or phrases in the word bank. Some may be chosen more than once. Change the forms if it's necessary.*

diatomic	aluminum oxide	anion	atoms	season
cation	covalent bond	electrons	valence	ionic bond
notation	Ne	negative	reactant	positive
share				

1) In the chemical symbol Fe_2O_3, the Fe^{3+} and the O^{2-} are connected through the _____.

2) In H_2, the two Hydrogen atoms are connected through the _____.

3) In balancing an equation, the number of the same _____ should be the same.

4) When one atom wants to give two electrons and the other atom wants to receive two electrons, they will be connected through an _____.

5) In a molecule composed of two identical atoms, such as Cl_2, they tend to _____ a certain number of electrons with each other.

6) If an atom loses two electrons, it will carry _____ charge. If an atom takes one electron, it will carry _____ charge.

7) The third bond among different atoms belongs to the metallic bond, which means that all the atoms will put their _____ into a sea of electrons.

8) If some aluminum reacts with oxygen sufficiently, it will produce _____.

9) In the family of noble gases, there are He, _____, Ar, Kr, Xe and etc.

10) In the chemical symbol H_2O, H will become a _____ and O will become an _____.

3. *Please balance the following chemical equations.*

1) $Fe_2O_3 + HCl = FeCl_3 + H_2O$

2) $Fe_2O_3 + H_2SO_4 = Fe_2(SO_4)_3 + H_2O$

3) $C + Fe_2O_3 = Fe + CO_2\uparrow$

4) $P + O_2 = P_2O_5$

5) $CO + Fe_3O_4 = Fe + CO_2$

6) $CO + Fe_2O_3 = Fe + CO_2$

7) $C_2H_5OH + O_2 = CO_2 + H_2O$

8) $KClO_3 = KCl + O_2\uparrow$

9) $Al + H_2SO_4 = Al_2(SO_4)_3 + H_2\uparrow$

10) $Al + HCl = AlCl_3 + H_2\uparrow$

4. *Translate the following sentences into English.*

1) 你无法改变食盐氯化钠分子中氯原子和钠原子的相对比率。

_____

2) 如果我们开始讨论有机化学,你就会有一些原子,很多碳原子和氢原子等等,它们组合在一起就会构成蛋白质。

_____

3) 一个原子想释放电子,因为它想在最外电子层实现稳定状态。

_____

4) 氧气很嫉妒惰性气体,因为惰性气体原子的最外层有八个电子。

_____

5) 我在这里没有讨论惰性气体的原因是它们没有形成共价键。

_____

_____

5. *Please write an essay of about* 120 *words on the topic* "***If We Don't Know Anything about Chemical Reactions...*** " *Some specific examples will be highly appreciated and watch out the spelling of some specialized terms you have learned in this unit.*

_____

_____

_____

_____

_____

_____

_____

_____

_____

_____

_____

_____

_____

_____

VI. Additional Reading

Brief Introduction on Dow Chemical Company

The Dow Chemical Company, commonly referred to as Dow, is an American multinational chemical corporation headquartered in Midland, Michigan, the United States, and the predecessor of the merged company DowDuPont. In 2017, it was the second largest chemical manufacturer in the world by revenue (after BASF) and as of February 2009, the third largest chemical company in the world by market capitalization (after BASF and DuPont). It ranked second in the world by chemical production in 2014.

Dow manufactures plastics, chemicals, and agricultural products. With a presence in about 160 countries, it employs about 54 000 people worldwide. The company has seven different major operating segments, with a wide variety of products made by each one. Dow's 2012 sales totaled approximately \$57 billion.

Dow has been called the "chemical companies' chemical company" in that most of its sales are to other industries rather than end users. Dow sells directly to end users primarily in the human and animal health and consumer products markets.

Dow is a member of the American Chemistry Council. The company tagline (宣传词) is "Solutionism".

On September 1, 2017, it merged with DuPont to create DowDuPont, the world's largest chemical company in terms of sales. In March 2018, it was announced that Jeff Fettig will become executive chairman of DowDuPont on July 1, 2018, and Jim Fitterling will become CEO of Dow Chemical on April 1, 2018.

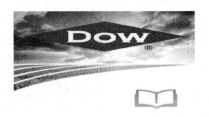

Dow is a large producer of plastics, including polystyrene(聚苯乙烯), polyurethane (聚氨酯), polyethylene(环聚乙烯), polypropylene(聚丙烯), and synthetic(合成的) rubber. It is also a major producer of ethylene oxide(氧乙烷), various acrylates(丙烯酸盐), surfactants(表面活性剂), and cellulose resins(纤维素树脂). It produces agricultural chemicals including the pesticide (杀虫剂) Lorsban and consumer products including Styrofoam. Some Dow consumer products including Saran wrap, Ziploc bags and Scrubbing Bubbles were sold to S. C. Johnson & Son in 1997.

Performance plastics

Performance plastics make up 25 percent of Dow's sales, with many products designed for the automotive and construction industries. The plastics include polyolefins (聚烯烃) such as polyethylene and polypropylene, as well as polystyrene used to produce Styrofoam(泡沫聚苯乙烯)insulating(绝缘的) material. Dow manufactures epoxy resin(环氧树脂)intermediates(中介) including bisphenol(双酚) A and epichlorohydrin(环氧氯丙烷). Saran resins and films are based on polyvinylidene chloride(PVDC, 聚偏二氯乙烯).

Performance chemicals

The Performance Chemicals (17 percent of sales) segment produces chemicals and materials for water purification(提纯), pharmaceuticals(医药品), paper coatings(涂层), paints and advanced electronics. Major product lines include nitroparaffins(硝基烷), such as nitromethane (硝基甲烷), used in the pharmaceutical industry and manufactured by Angus Chemical Company, a wholly owned subsidiary of the Dow Chemical Company. Important polymers include

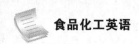

Dowex ion exchange resins, acrylic (丙烯酸的) and polystyrene latex (胶乳), as well as Carbowax polyethylene glycols(乙二醇). Specialty chemicals are used as starting materials for production of agrochemicals(农用化学品) and pharmaceuticals.

Water purification

Dow Water and Process Solutions (DW&PS) is a business unit which manufactures Filmtec reverse osmosis (渗透) membranes (膜) which are used to purify water for human use in the Middle East. The technology was used during the 2000 Summer Olympics and 2008 Summer Olympics.

Agricultural sciences

Agricultural Sciences, or Dow AgroSciences, provides 7 percent of sales and is responsible for a range of insecticides(杀虫剂)(such as Lorsban), herbicides(除草剂) and fungicides(杀真菌剂). Seeds from genetically modified plants are also an important area of growth for the company. Dow AgroSciences sells seeds commercially under the following brands：Mycogen [grain corn, silage(青贮饲料) corn, sunflowers, alfalfa(苜蓿), and sorghum(高粱)], Atlas (soybean), PhytoGen (cotton) and Hyland Seeds in Canada [corn, soybean, alfalfa, navy beans (菜豆) and wheat].

Basic plastics

Basic plastics (26 percent of sales) end up in everything from diaper liners to beverage bottles and oil tanks. Products are based on the three major polyolefins—polystyrene (such as Styron resins), polyethylene and polypropylene.

Basic chemicals

Basic chemicals (12 percent of sales) are used internally by Dow as raw materials and are also sold worldwide. Markets include dry cleaning, paints and coatings, snow and ice control and the food industry. Major products include ethylene glycol, caustic(腐蚀性的) soda, chlorine, and vinyl chloride monomer(氯乙烯单体) (VCM, for making PVC). Ethylene oxide and propylene oxide(环氧丙烷) and the derived alcohols ethylene glycol and propylene glycol(丙二醇) are major feedstocks for the manufacture of plastics such as polyurethane and PET.

Hydrocarbons and energy

The Hydrocarbons and Energy operating segment (13 percent of sales) oversees energy management at Dow. Fuels and oil-based raw materials are also procured(取得). Major feedstocks for Dow are provided by this group, including ethylene, propylene(丙烯), butadiene (丁二烯), benzene and styrene(苯乙烯).

In 2003, Dow agreed to pay $2 million, the largest penalty ever in a pesticide case, to the state of New York for making illegal safety claims related to its pesticides. The New York Attorney General's Office stated that Dow AgroSciences had violated a 1994 agreement with the State of New York to stop advertisements making safety claims about its pesticide products. Dow stated that it was not admitting to any wrongdoing, and that it was agreeing to the settlement to avoid a costly court battle.

According to the United States Environmental Protection Agency (EPA), Dow has some responsibility for 96 of the United States' Superfund toxic waste sites, placing it in 10th place by number of sites. One of these, a former UCC uranium(铀) and vanadium(钒) processing facility near Uravan, Colorado, is listed as the sole responsibility of Dow. The rest are shared with numerous other companies. Fifteen sites have been listed by the EPA as finalized (cleaned up) and 69 are listed as "construction complete", meaning that all required plans and equipment for cleanup are in place.

In 2007, the chemical industry trade association—the American Chemical Council—gave Dow an award of "Exceptional Merit" in recognition of longstanding energy efficiency and conservation efforts. Between 1995 and 2005, Dow reduced energy intensity (BTU per pound produced) by 22 percent. This is equivalent to saving enough electricity to power eight million US homes for a year. The same year, Dow subsidiary, Dow AgroSciences, won a United Nations Montreal Protocol Innovators Award for its efforts in helping replace methyl bromide(溴甲烷)—a compound identified as contributing to the depletion of the ozone layer(臭氧层). In addition, Dow AgroSciences won an EPA "Best of the Best" Stratospheric Ozone Protection Award. The United States Environmental Protection Agency named Dow as a 2008 Energy Star Partner of the Year for excellence in energy management and reductions in greenhouse gas emissions. In July 2010, Dow became a worldwide partner of the Olympic Games. The sponsorship extends to 2020.

In September 2004, Dow obtained the naming rights to the Saginaw County Event Center in Saginaw, Michigan; the center is now called the Dow Event Center. The Saginaw Spirit (of the Ontario Hockey League) plays at the center, which also hosts events such as professional wrestling and live theater.

In October 2006, Dow bought the naming rights to the stadium used by the Great Lakes Loons, a Single-Aminor league baseball team located in its hometown of Midland, Michigan. The stadium is called Dow Diamond. The Dow Foundation played a key role in bringing the Loons to the city.

In 2010, Dow signed a $100 million (£63 million) 10-year deal with the International Olympic Committee and agreed to sponsor the £7 million Olympic Stadium wrap. Dow also sponsors Austin Dillon's #3 Chevrolet(雪佛兰牌汽车) in the Monster Energy NASCAR Cup Series.

Lab Safety Academy

On May 20, 2013, Dow launched the Dow Lab Safety Academy, a website that includes a large collection of informational videos and resources that demonstrate best practices in laboratory safety. The goal of the website is to improve awareness of safety practices in academic research laboratories and to help the future chemical workforce develop a safety mindset. As such, the Dow Lab Safety Academy is primarily geared toward university students. However, Dow has made the content open to all, including those already employed in the chemical industry. The Dow Lab Safety Academy is also available through the Safety and Chemical Engineering Education program, an affiliate of American Institute of Chemical Engineers (AIChE); and The Campbell Institute, an organization focusing on environment, health and safety practices.

The Dow Lab Safety Academy is one component of Dow's larger laboratory safety initiative launched in early 2012, following a report from the U.S. Chemical Safety Board that highlighted the potential hazards associated with conducting research at chemical laboratories in academic institutions. Seeking to share industry best practices with academia, Dow partnered with several U.S. research universities to improve safety awareness and practices in the departments of chemistry, chemical engineering, engineering and materials. Through the pilot programs with U.C. Santa Barbara (UCSB), University of Minnesota, and Pennsylvania State University, Dow worked with graduate students and faculty to identify areas of improvement and develop a culture of laboratory safety.

Nature Conservancy

In January 2011, The Nature Conservancy(保护) and the Dow Chemical Company announced a collaboration to integrate the value of nature into business decision-making. Scientists, engineers, and economists from The Nature Conservancy and Dow are working together at three pilot sites (North America, Latin America, and TBD) to implement and refine models that support corporate decision-making related to the value and resources nature provides. Those ecosystem services include water, land, air, oceans and a variety of plant and animal life. These sites will serve as a "living laboratories", to validate(使合法化) and test methods and models so they can be used to inform more sustainable business decisions at Dow and hopefully influence the decision-making and business practices of other companies.

(*If you want to find more information about this corporation, please log on https://en.wikipedia. org/wiki/Dow_Chemical_Company*)

1. *Read the passage quickly by using the skills of skimming and scanning. And choose the best answer to the following questions.*

    1) The Dow Chemical Company is headquartered in _____.

        A. Midland   B. Michigan

        C. the United States   D. the United Kingdom

2) Which is not one of Dow's products?

    A. Plastics.   B. Agricultural products.

    C. Chemicals.   D. Snacks.

3) The plastics include polyolefins such as _____ used to produce Styrofoam insulating material.

    A. polyethylane, polypropylene, polystyrene

    B. polyethylene, polystyreene, polypropylene

    C. polystyrene, polyethylene, polypropylene

    D. polystyrene, polyethylene, polypropyleane

4) Which is not one of the important polymers?

    A. Dowex ion change resins.   B. Dowex ion exchange resins.

    C. Acrylic and polystyrene latex.   D. Carbowax polyethylene glycols.

5) In 2007, the chemical industry trade association—the American Chemical Council—gave Dow an award of "Exceptional Merit" in recognition of longstanding _____.

    A. energy effectiveness and conservation efforts

    B. energy efficiency and conservation efforts

    C. energy efficiency and conversation efforts

    D. energy effectiveness and converse efforts

6) In 2010, Dow signed a _____ deal with the International Olympic Committee and agreed to sponsor the _____ Olympic Stadium wrap.

    A. $100 million, 9-year, £7 million   B. $100 million, 10-year, £7 million

    C. $100 million, 11-year, £8 million   D. $100 million, 12-year, £8 million

7) Who are the Dow Lab Safety Academy's target group?

    A. University students.   B. High school students.

    C. Middle school students.   D. Primary school students.

8) Dow partnered with several U. S. research universities to improve _____ in the departments of chemistry, chemical engineering, engineering and materials.

    A. safety practices   B. safety awareness

    C. safety awareness and practices   D. safety measures

9) In January 2011, what do The Nature Conservancy and the Dow Chemical Company decide to do?

    A. To cooperate to integrate the value of nature into business decision-making.

    B. To cooperate to integrate the value of society into business decision-making.

    C. To cooperate to integrate the value of nature into high-level decision-making.

    D. To cooperate to integrate the value of society into high-level decision-making.

10) Which is not one of the three pilot sites where scientists, engineers, and economists from The Nature Conservancy and Dow are working together?

    A. North America.   B. TBD.

        C. Mexico.   D. Latin America.

2. *In this part , the students are required to make an oral presentation on either of the following topics.*

   1) The secrets of Dow's success.

   2) The prospects for Dow.

习题答案

Unit Four Food Processing

I. Pre-class Activity

Directions: *Please read the general introduction about Friedrich Christian Accum and share your knowledge on the food processing technologies with your classmates. (For more information at http://www.chemeurope.com/en/encyclopedia/Friedrich_Accum.html)*

Friedrich Christian Accum

Friedrich Christian Accum (29 March 1769, Bückeburg – 28 June 1838, Berlin) was a German chemist, whose most important achievements were in the areas of gaslight, the fight against poisonous foods, and the popularizing of chemistry.

Accum lived in London from 1793 to 1821, where, as a self-employed chemist, he manufactured and sold chemical and scientific instruments, gave fee-based public lectures in practical chemistry, and collaborated with various research institutions.

In 1820, Accum published *A Treatise on Adulterations of Food*, in which he denounced the use of poisoned foodstuffs. The work marked the beginning of a concern with nutrition. Accum was the first person to tackle the subject and to bring a wide audience with his work. Although his book sold extremely well, his attempts to raise public awareness in these areas made him many enemies among London foodstuff purveyors. Accum left England after a lawsuit was brought against him. He lived out the rest of his life as a teacher at an industrial institution in Berlin. His publications, most of which were written in English, are in a style accessible to the general public of the period. Accum thus made important contributions to the popularization of chemistry in the period.

II. Specialized Terms

Directions：*Please memorize the following specialized terms before the class so that you will be able to better cope with the coming tasks.*

acid n.酸;酸味物质

activator n.激活剂,活化剂

additive action 相加作用

age v.[化]使熟化,使老化

air sifter with pneumatic conveyor 气动风筛机

alternative n.替代品

antibiotic n.抗生素

antibiotic-associated adj.与抗生素相关的

assembly line（工厂产品）装配线

auger n.螺旋钻

blade n.(机器上) 叶片;桨叶

calorimeter n.测热计,热量计

coagulate v.凝结,使凝结

conching n.混合搅拌

conching machine（巧克力配料）揉搅机,精炼机

conical adj.圆锥形的

consistency n.稠度

controlled atmosphere（CA）storage 气控贮藏

convey v.传运 ,输送

conveyor n.输送机,传送带

cork n.软木塞

counter rotating gears 反旋转齿轮

counterpart n.对应物

cure v.加工处理

demijohn n.细颈瓶,小口大肚瓶

denature v.使改变本性

double-fermented adj.双重发酵的

eliminate v.除去;排除

emulsion n.乳剂,乳状液

enrich v.(食品)增补(营养素),强化;富集;加料

enriched food 强化(营养素)食品

enricher n.强化剂

extract v.提取,榨出

extracted adj.提取的

fermentation locks 发酵气泡阀

fermenter n.发酵罐

filtration equipment 过滤设备

finished food 制成的食品

flavology n.(食品)风味学

flavoring n.调味品

flavor additive 香味添加剂

flavor adjunct 食品香精辅助物

food color 食品(用)色素

food dye 食品(用)色素

gas storage 气体保鲜法

gear n.齿轮;(排)挡

GM food（genetically modified food）转基因食品

grind v.摩擦,磨碎

grinding n.碾磨

hold v.静放

homogenize v.使……均质,使均匀

hopper n.漏斗,料斗

horizontal mixer 卧式混合机

inactivated adj.未激活的

instantizer n.速溶剂;速溶机

irradiation n.辐射,照射;食品照射保藏法

isolate flavor 单离食品香料

jot n.添加剂

juice extractor 榨汁机

juicer n.榨汁器

knead v.揉捏,揉合

leaven v.使(面团)发酵

leavening agent 发酵剂,酵母

massage v.揉

masticating juicer 咀嚼式榨汁机

melangeur n.(糖果食品工业中用的)搅拌器

mellow v.使成熟,使柔和

microwaveable food 可微波食品

milk substitute 代乳食品

minimally processed foods 最少加工食品

modification n.更改,改变

modified atmosphere(MA)storage 变气贮藏

mould n.模子,铸模(=mold)

obstruction n.障碍物,妨碍物

outlet n.排放管

paste n.糊,浆糊

pasteurize v.用巴氏灭菌法对(牛奶等)消毒(灭菌)

phytochemicals n.植物化学物质

plastic funnel 塑料漏斗

pneumatic adj.气动的,风动的

press cake 滤饼,压榨饼

processed food 加工食品

processing plant 食品加工厂

profile v.异型,仿型

proofer n.(有控制发酵温度的酵母发酵)面团贮存器

provision n.供应,食品

pulverize v.磨成粉,粉碎

rack n.支架,架子

rattle v.发出嘎嘎声

reamer n.铰刀

recall v.召回,收回

reconstitute v.(加水使脱水食物)复原

refine v.精炼,提炼

screw n.螺旋,螺纹

seal n.密封

secondary product 副产品

semimoist food 半(湿)食品

separator n.分离器

shooter n.滑槽

sieve n.筛,漏勺

sifter n.筛子,筛粉机

siphon v.用虹吸管吸出,抽取

siphon tube 虹吸管

soluble adj.可溶的,易溶解的

spray drying 喷雾干燥

stability n.稳定性,耐贮性

stabilizer n.稳定剂

starter culture 起子培养

sterilize v.杀菌,消毒

III. Watching and Listening

Task One   A Journey in a Chocolate Factory

New Words

raw adj.生的,未加工的

cocoa n.可可,可可粉

Kit Kat 奇巧巧克力

staggering adj.难以置信的

maintenance n.维持,保持

unload v.卸货;去除负担

hook v.用钩挂

avalanche n.雪崩

视频链接及文本

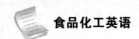

grubby adj.污秽的,邋遢的

pebble n.卵石

literally adv. 确实地,真正地;[口]简直

Ivory Coast 象牙海岸(非洲)

grit n.细沙,沙砾

crack v.破裂,打开

hitch v.(免费)搭乘他人之车

de-shelling 去壳,去皮

crush v.压破,压碎,镇压;弄皱;挤榨,榨出

nib n.可可豆的碎粒

the Holy Grail n. [宗]圣杯,圣盘

vacuum v.用真空吸尘器清扫

recipe n.食谱;秘诀

pipe v.以管输送

stainless adj.不锈的;不会脏的

steel n.钢,钢铁

vessel n.容器;大船

component n.组分;零件

coarse adj.粗鄙的;粗糙的

churn out 快速生产;大量生产

blender n.掺和机,搅拌器

conch n.贝壳,海螺壳

Exercises

1. *Watch the video for the first time and choose the best answers to the following questions.*

   1) Eight hundred production staff members work around the clock to produce millions of bars to the consumers in _____.

      A. the UK and Ireland　　　　　B. Canada

      C. America　　　　　　　　　　D. Europe

   2) According to the video clip, the British people eat a staggering _____ chocolate bars every single minute.

      A. 1670　　　　　　　　　　　B. 2000

      C. 1900　　　　　　　　　　　D. 2200

   3) Ninety percent of British chocolate begins life as cocoa beans in _____.

      A. Brazil　　　　　　　　　　　B. West Africa

      C. Argentina　　　　　　　　　D. Southern Chile

   4) The _____ of the bean is the Holy Grail of chocolate making.

      A. color　　　　　　　　　　　B. smell

      C. shell　　　　　　　　　　　D. nib in the center

   5) Sean Conricode is the factory's chocolate specialist as well as _____ for over thirty-eight years.

      A. the accountant

      B. the keeper of recipes

      C. an employer from the chocolate production line

      D. the maintenance worker

2. *Watch the video again and decide whether the following statements are true or false.*

   1) It takes raw cocoa beans just 24 hours from arriving by truck to leaving later as chocolate bars ready for the supermarket.(　)

   2) Cocoa bean expert Steve Calpin has been working in the factory for over fifty years.

( )

3) The main harvest of pods from cocoa trees happens in May.( )

4) The workers used to find shoes, snakes and coins in tons of beans.( )

5) To crack their hard outer shell and release the cocoa-rich center, the beans do not need to be cleaned.( )

3. *Watch the video for the third time and fill in the following blanks.*

After an hour, the roasted nibs are fed into a giant food 1)_____, which transforms them into a 2)_____ brown 3)_____ called Liquor.

"The thing that's made it liquid like that is the fact that's come out of the nib."

"The same way that 4)_____ comes out of an 5)_____, your fats coming out of the chocolate nib?"

"That's correct."

"Have you added anything at all to this nib?"

"Just a little bit of love and care and 6)_____."

"We all need 7)_____ of that, mate."

"I know."

Eight tons of 8)_____ liquor is now piped from the bean processing plant to the factory next door. Here they will turn the bitter thick liquid into 9)_____ tons of sweet milk 10)_____.

4. *Look at the following pictures of chocolate-making, and try to describe it to your partner with the information provided in the video.*

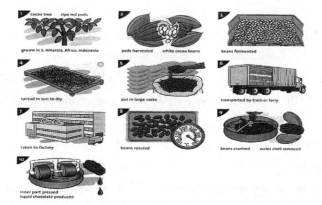

Task Two   Sweet Attacks

New Words

gourmet n.美食家,讲究吃喝的人

snap n.(树枝等的)突然折断

gooey adj.黏软的,感伤的

sophistication n.老练,精明

视频链接及文本

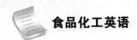

daiquiri n.台克利(鸡尾)酒

drastically adv.大大地,彻底地;激烈地

deposit v.放置 n.[医]沉积物

core n.中心,核心;果心,果核

tray n.盘子;托盘

mogul n.有权势的人,大人物;〈美〉富豪

slurry n.泥浆,浆料

engrossing adj.趣味无穷,引人入胜的

rotate v.(使某物)旋转

confectioner n.糖果制造人,糖果店

glaze v.使光滑

tumble v.翻来覆去,翻腾

friction n.摩擦;冲突,不和

inexhaustible adj.无穷无尽的,用不完的

craving n.渴望,渴求

Exercises

1. *Watch the video for the first time and choose the best answers to the following questions.*

1) Every day the Jelly Belly factory in Fairfield, California, produces _____ pounds of its famous gourmet beans.

   A. 230 000   B. 250 000

   C. 260 000   D. 270 000

2) Jelly Belly produces fifty official flavors among which _____ is not mentioned in this vedio clip.

   A. Very Cherry   B. Strawberry Daiquiri

   C. Butter Popcorn   D. Brandy

3) The female owner of Jelly Belly says that the taste testing will rate each Jelly Belly flavor on the following aspects except _____.

   A. nutrient   B. color

   C. texture   D. flavor

4) When the Jelly Belly's tender core is created, corn syrup is used to _____.

   A. sweeten   B. harden

   C. thicken   D. lighten

5) If the Jelly Belly won't be too soft or too hard, one of the workers says that's a factor of _____.

   A. concentration   B. time

   C. temperature   D. ingredient

2. *Watch the video again and decide whether the following statements are true or false.*

1) Their latest creations include smoothie flavors and sports beans for athletes who need an extra energy boost.( )

2) It takes less than a week to make each Jelly Belly.( )

3) During the wedding part, Jelly Belly uses the automated systems because it makes the better quality that we want to have made today.( )

4) The final part of making Jelly Belly is to polish it inside these giant steel drums.( )

5) The video believe Jelly Belly will likely flourish for at least another century.( )

3. *Watch the video for the third time and fill in the following blanks.*

Walt Wieser is supervisor of the Jelly Belly factory. With nearly twenty years of 1)
_____ on the floor, he's one of the most 2)_____ candy-makers. "Right now, you're in
the moulding room, the second step to jelly." The candy 3)_____ travels down stairs
through a tube to the moulding 4)_____, the powerhouse of the factory. It's a multi-tasker
that molds, 5)_____ and handles up to 6)_____ pounds of slurry per hour. Without this
machine, 7)_____ would be drastically altered. The slurry is deposited into starch moulds
that 8)_____ the bean core. Each tray holds twelve hundred sixty bean moles, and the
mogul must run about twenty eight trays every minute to meet the daily 9)_____. The
mogul cooks the slurry mixture at exactly two hundred fifty 10)_____.

4. *Share your opinions with your partners on the following topics for discussion.*

1) Examples of the benefits of modern technology in improving food processing.

2) The reasons and motivations why workers stay in a company for over 20 years.

IV. Talking

Task One   Classical Sentences

Directions: *In this section, some popular sentences are supplied for you to read and to memorize.
Then, you are required to simulate and produce your own sentences with reference to the structure.*

General Sentences

1. Do you really want to know what I think?
   你真想知道我在想什么吗?

2. Please give me your frank opinion.
   请告诉我你真实的想法。

3. Of course I want to know what your opinion is.
   当然,我很想知道你的看法。

4. What do you think? Is that right?
   你认为怎样? 可以吗?

5. Certainly. You're absolutely right about that.
   当然,你完全正确。

6. I think you're mistaken about that.
   我想你弄错了。

7. I'm anxious to know what your decision is.
   我很想知道你的决定。

8. That's a good/great/fantastic/excellent idea.
   这个想法很好。

9. In my opinion, that's an excellent idea.
   我认为这是个好主意。

10. I'm confident you've made the right choice.
   我相信你做了个正确的决定。

11. I want to persuade you to change your mind.
   我想劝你改变主意。

12. Will you accept my advice?
   你会接受我的建议吗?

13. He didn't want to say anything to influence my decision.
   他不想说任何话来影响我做决定。

14. She refuses to make up her mind.
   她不肯下决心。

15. I assume you've decided against buying a new car.
   我想你已经决定不买新车了吧。

16. It took him a long time to make up his mind.
   他用了很久才下定决心。

17. You have your point of view, and I have mine.
   你有你的观点,我也有我的想法。

18. You approach it in a different way from mine.
   你处理这件事的方式和我不一样。

19. I won't argue with you, but I think you're being unfair.
   我不想和你争辩,但是我认为你这样不公平。

20. That's a liberal point of view.
   那是一种自由主义的观点。

21. He seems to have a lot of strange ideas.
   他好像有很多奇怪的想法。

22. I don't see any point in discussing the question any further.
   我看不出有任何必要进一步讨论这个问题。

23. What alternatives do I have?
   我还有什么方法?

24. Everyone is entitled to his own opinion.
   每个人都有自己的观点。

25. She doesn't like anything I do or say.
   无论我做什么说什么,她都不喜欢。

26. There are always two sides to everything.
   任何事物都有两面性。

27. We have opposite views on this.
   关于这个问题,我们有不同的观点。

28. Please forgive me. I didn't mean to start an argument.
请原谅,我并不想引起争论。

29. I must know your opinion. Do you agree with me?
我必须了解你的想法。你同意吗?

30. What points are you trying to make?
你想表达什么观点?

31. Our views are not so far apart, after all.
毕竟我们的意见没有多大分歧。

32. We should be able to resolve our differences.
我们应该能解决我们的分歧。

33. If you want my advice, I don't think you should go.
如果你想征求我的意见,我认为你不应该去。

34. I suggest that you tear up the letter and start over again.
我建议你把信撕掉,再重新写一遍。

35. It's only a piece of suggestion, and you can do what you please.
这只是一个建议,你可以做你想做的。

36. Let me give you a little fatherly advice.
让我给你提点慈父般的建议。

37. If you don't like it, I wish you would say so.
如果你不喜欢,我希望你能说出来。

38. Please don't take offense. I only wanted to tell you what I think.
请别生气,我只是想告诉你我的想法。

39. My feeling is that you ought to stay home tonight.
我觉得你今晚应该待在家里。

40. It's none of my business, but I think you ought to work harder.
这不关我的事,但我认为你应该更努力地工作。

41. In general, my reaction is favorable.
总体来说,我是赞成的。

42. If you don't take my advice, you'll be sorry.
如果你不听我的劝告,你会后悔的。

43. I've always tried not to interfere in your affairs.
我总是尽量不干涉你的事情。

44. I'm old enough to make up my own mind.
我已经长大了,可以自己做决定了。

45. Thanks for the advice, but this is something I have to figure out myself.
谢谢你的建议,但这是我必须自己解决的问题。

46. He won't pay attention to anybody. You're just wasting your breath.
他谁的建议都不会听,你们是在白费口舌。

47. You can go whenever you wish.

你愿意什么时候去就什么时候去。

48. We're willing to accept your plan.

我们愿意接受你的计划。

49. He knows it's inconvenient, but he wants to go anyway.

他知道不方便,但他无论如何都想走。

50. He insists that it doesn't make any difference to him.

他坚持说这对他没有任何影响。

Specialized Sentences

1. Food processing is the transformation of agricultural products into food, or of one form of food into other forms.

食品加工是指将农产品转化为食品,或将一种形式的食品转化为其他形式。

2. Food-centric technological inventions are popping up around the globe to improve the food industry.

以食品为中心的技术发明在全球各地涌现,以促进食品工业的发展。

3. The idea of 3D printed food isn't exactly mouth-watering.

3D 打印食物的想法并不令人垂涎欲滴。

4. The technology of 3D printed food stands to disrupt the food industry on at least some level.

3D 打印食品技术可以在某种程度上颠覆食品行业。

5. Food processing includes many forms of processing foods, from grinding grain to make raw flour to home cooking to complex industrial methods used to make convenience foods.

食品加工包括多种形式的加工食品,包括从将谷物磨成生面粉到家庭烹饪,再到用于制作方便食品的复杂工业方法。

6. Primary food processing is necessary to make most foods edible.

要使大多数食物可以食用,初级食品加工是必需的。

7. Primary food processing includes ancient processes such as drying, threshing, winnowing, and milling grain.

初级食品加工包括古老的工艺,如干燥、脱粒、风选和碾磨谷物。

8. Primary food processing also includes deboning and cutting meat, freezing and smoking fish and meat, extracting and filtering oils.

初级食品加工还包括去骨、分割肉类,冷冻和熏制鱼类食品、肉类食品,提取和过滤油脂。

9. Contamination and spoilage problems in primary food processing can lead to significant public health threats.

初级食品加工中的污染和变质问题可能导致严重的大众健康威胁。

10. Secondary food processing turns the ingredients into familiar foods.

二级食品加工将配料变成熟悉的食物。

11. Baking bread is an example of secondary food processing.

烘烤面包是二次食品加工的一个例子。

12. Secondary food processing is the everyday process of creating food from ingredients that are ready to use.

二级食品加工是用即食原料制作食品的日常过程。

13. Fermenting fish and making wine, beer, and other alcoholic products are traditional forms of secondary food processing.

发酵鱼, 酿造葡萄酒、啤酒和其他酒类产品是传统形式的二次食品加工。

14. Sausages are also a common form of secondary processed meat, which has already undergone primary processing.

香肠是对肉进行初级加工后, 再进行二级加工的常见形式。

15. Tertiary food processing is the commercial production of what is commonly called processed food.

第三级食品加工是通常所说的对加工食品进行的商业化生产。

16. Processed food includes ready-to-eat or heat-and-serve foods, such as TV dinners and reheated airline meals.

加工食品包括即食食品或热食食品, 如看电视时的佐餐晚餐和再加热的航空餐。

17. Tertiary food processing has been criticized for promoting over-nutrition and obesity, containing too much salt, too little fiber, and otherwise being unhealthful.

三级食品加工因为营养过量和肥胖症, 含有过多的盐和过少的纤维, 以及其他方面的不健康原因而受到批评。

18. Food processing dates back to the prehistoric ages.

食品加工可以追溯到史前时代。

19. Basic food processing involved chemical enzymatic changes to the basic structure of food in its natural form.

基本食品加工涉及以化学酶对天然食品基本结构的改变。

20. Salt-preservation was especially common for foods that constituted warriors and sailors' diets until the introduction of canning methods.

在采用罐装方法之前, 用盐来保存战士和海员的食物是特别常见的方法。

21. Modern food processing technology was developed in a large part to serve military needs.

现代食品加工技术在很大程度上是为了满足军事需求而开发的。

22. By the 20th century, automatic appliances like microwave oven, blender, and rotimatic paved way for convenience cooking.

到了 20 世纪, 自动化设备, 如微波炉、搅拌机和自动烙饼机, 有助于便捷烹饪。

23. In the second half of the 20th century, food processing companies marketed their products especially towards middle-class working wives and mothers.

在 20 世纪下半叶, 食品加工公司特别针对中产阶级的职业妻子和母亲销售其产品。

24. Benefits of food processing include toxin removal, preservation, easing marketing and

distribution tasks.

食品加工的好处包括可以去除毒素、延长食物保存期、简化营销和分销工作等。

25. Food processing enables transportation of delicate perishable foods across long distances.

食品加工可以长距离运输精致易腐食品。

26. Modern supermarkets would not exist without modern food processing techniques, and long voyages would not be possible.

没有现代食品加工技术，现代超市就不会存在，长途航行也是不可能的。

27. Processed foods are usually less susceptible to early spoilage than fresh foods.

加工食品通常比新鲜食品更不易受早期腐烂的影响。

28. Processed foods are better suited for long-distance transportation from the source to the consumer.

加工食品更适合从源产地到消费者的长途运输。

29. When they were first introduced, some processed foods helped to alleviate food shortages and improved the overall nutrition of populations.

当它们首次引入市场时，某些加工食品有助于缓解粮食短缺并改善人口的总体营养。

30. Processing can also reduce the incidence of foodborne disease.

加工也可以减少食源性疾病的发病率。

31. Fresh materials, such as fresh produce and raw meats, are more likely to harbor pathogenic microorganisms.

新鲜农产品和生肉等生鲜材料更容易携带病原微生物。

32. The act of processing can often improve the taste of food significantly.

加工过程通常可以显著改善食物的味道。

33. Mass production of food is much cheaper overall than individual production of meals from raw ingredients.

食品的大规模生产总体上来说比由原料单次制做食品便宜得多。

34. A large profit potential exists for the manufacturers and suppliers of processed food products.

加工食品的制造商和供应商拥有巨大的利润空间。

35. Processed food freed people from the large amount of time involved in preparing and cooking "natural" unprocessed foods.

加工食品使人们免于为了准备和烹饪"天然"未加工食品耗费大量时间。

36. The increase in free time from enjoying processed food allows people much more choice in life style than previously allowed.

由于加工食品带来的空闲时间的增加，让人们的生活方式比以前有了更多选择。

37. Fully prepared ready meals can be heated up in the microwave oven within a few minutes.

即食食品在几分钟内可以在微波炉中完成加热。

38. Modern food processing also improves the quality of life for people with allergies and

diabetics.

现代食品加工也改善了过敏症和糖尿病患者的生活质量。

39. During food processing, extra nutrients such as vitamins can also be added, which are not contained naturally.

在食品加工过程中,还可以添加额外的营养素,如并非天然含有的维生素。

40. Processing of food can decrease its nutritional density.

加工食品可降低其营养密度。

41. The amount of nutrients lost depends on the food and processing method.

营养素损失数量取决于食物和加工方法。

42. Except for the nutrition loss, using some food additives represents another safety concern.

除了营养损失外,某些食品添加剂的使用是另一个安全问题。

43. Food processing is typically a mechanical process which can introduce a number of contamination risks.

食品加工是一种典型的可能会带来许多污染风险的机械生产过程。

44. These days, acceptance of food products very often depends on potential benefits and risks associated with the food.

当前,对食品的接受往往取决于与食品有关的潜在利益和风险。

45. One of the main sources of sodium in the diet is processed foods.

饮食中钠的主要来源之一是加工食品。

46. Processed foods may actually take less energy to digest than natural foods.

与天然食物相比,消化加工食品实际上所需的能量更少。

47. Foods that have undergone processing, including some commercial baked goods, desserts and microwave popcorn, sometimes contain trans fats.

经过加工的食品,包括一些商业烘焙食品、甜点和微波炉爆米花,有时会含有反式脂肪。

48. Fast food is a mass-produced food which is typically prepared and served quicker than traditional foods.

快餐是一种大规模生产的食品,通常比传统食品更容易制作和提供。

49. The concept of processed food for sale is closely connected with urban developments.

销售加工食品的理念与城市发展密切相关。

50. Similarly to the advances made in other areas of our lives, food is undergoing its own revolution.

与我们生活中其他方面不断进步相似,食品也正在经历自己的革命。

Task Two   Sample Dialogue

Directions: *In this section, you are going to read several times the following sample dialogue about the relevant topic. Please pay special attention to five C's (culture, context, coherence, cohesion and critique) in the dialogue and get ready for a smooth communication in the coming task.*

Friends talking about recent visit at a food factory

Molly: Hey, guys. I heard that you paid a visit to a food factory recently. Where have you been?

Peter: We have visited a food company making canned food. It was an interesting experience which really opened my eyes to food processing.

George: Yeah, especially the assembly line. If you have time, it will deserve a visit.

Molly: The assembly line? Is it efficient?

George: During the whole process, the fruits are delivered by the conveyer belt stably and quickly.

Peter: The workers just need to stand by and get rid of the unqualified products occasionally.

Molly: I'm quite interested in the sterilization process. How do they achieve it?

Peter: That's the core part. There are two steps. First, after washing and peeling, the fruits will bathe themselves in water of disinfected matters. Then, they will be boiled in hot water for 20 minutes.

George: Normally, a canned fruit can keep for 2 years.

Molly: It seems that not too much additives are used. So can it be classified as natural and healthy food?

George: Definitely not natural. Since it's produced for commercial purpose, it should be categorized as processed food.

Peter: And any individual should not have too much once a time, because it contains too much sugar in it.

Molly: Knowing something about the production seems really useful. Next time, when you are heading for next factory, take me with you.

George: No problem.

Task Three Simulation and Reproduction

Directions: *The class will be divided into three major groups, each of which will be assigned a topic to discuss with your partners. In the process of discussion, please observe the principles of cooperation, politeness and choice of words. One of the groups will be chosen to demonstrate the discussion to the class.*

1) More examples of primary, secondary and tertiary food processing.

2) Differences of cooking method between the Chinese people and the Americans.

3) How to cook your favorite dish.

Task Four Discussion and Debate

Directions: *The class will be divided into two groups. Please choose your stand in regard to the following controversy and support your opinions with scientific evidences. Please refer to the specialized terms and classical sentences in the previous parts of this unit.*

These days, it is quite common for consumers to purchase processed foods in the daily life. Some believe the edge of processed foods outweighs its disadvantage. However, multiple opposing opinions show up among the public who assume that its advantages are not worthy to be mentioned, compared with its cost. Are you a member of the former group or the latter one? Why?

V. After-class Exercises

1. *Match the English words in Column A with the Chinese meaning in Column B.*

A	B
1) ferment	a. 稠度
2) assembly line	b. 揉捏,揉合
3) reconstitute	c. (使)发酵
4) consistency	d. (加水使脱水食物)复原
5) knead	e. 使成熟,使柔和
6) denature	f. 蒸发
7) evaporation	g. 杀菌,消毒
8) mellow	h. (工厂产品的) 装配线
9) instantizer	i. 使……改变本性
10) sterilize	j. 速溶剂

2. *Judge whether the following statements about food processing are true or false and then write "T" or "F" in the bracket. If it's wrong, please correct it.*

1) Primary food processing is the commercial production of what is commonly called processed food. (   )

2) Primary food processing turns the ingredients into familiar foods. (   )

3) Sausage is an example of primary food processing. (   )

4) Tertiary food processing includes ancient processes such as drying, threshing, winnowing, and milling grain. (   )

5) Food processing makes it possible to transport delicate perishable foods across long distances all over the world. (   )

6) Tertiary food processing has been criticized for promoting over-nutrition and obesity. (   )

7) Modern food processing technology was developed mainly to serve commercial needs. (   )

8) Processed foods may actually take less energy to digest than 3D printed foods. (   )

9) Processing of food can decrease its nutritional density. (   )

10) In the second half of the 20th century, housewives were the main consumers of food. (   )

3. *Translate the following sentences into English.*

1）食品加工确实有一些好处，例如使食品保质期更长，使产品更方便享用。

_____

_____

2）热量会破坏维生素 C。因此，罐装水果比新鲜水果含有更少的维生素 C。

_____

_____

3）特定某种添加剂的健康风险因人而异，例如，使用糖作为添加剂危及糖尿病患者。

_____

_____

4）随着化学添加剂被人们熟知，法律和监管实践的演变试图使得这些加工食品更加安全。

_____

_____

5）我们的乡镇企业有木材加工、食品加工和饮食服务业。

_____

_____

4. *Please write an essay to describe the following pictures of the manufacturing of potato chips. Try to pay attention to all the provided details and add more information if you want. Please watch out the spelling of some specialized terms you have learnt in this unit.*

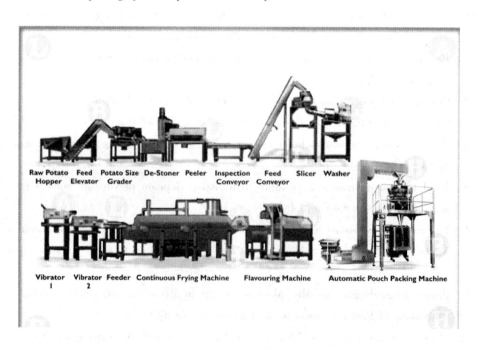

_____

_____

_____

_____

_____

_____

_____

_____

_____

_____

_____

_____

_____

_____

_____

_____

_____

_____

_____

VI. Additional Reading

A Profile of Heinz

The H. J. Heinz Company, or Heinz, is an American food processing company with world headquarters in Pittsburgh, Pennsylvania. Originally, the company was founded by Henry John Heinz in 1869. The H. J. Heinz Company manufactures thousands of food products in plants on six continents, and markets these products in more than 200 countries and territories. The company claims to have 150 number-one or number-two brands worldwide. Heinz ranked first in ketchup in the US with a market share in excess of 50%; the Ore-Ida label held 46% of the frozen potato sector in 2003.

On February 14, 2013, Heinz agreed to be purchased by Berkshire Hathaway and 3G Capital for $ 23 billion. On March 25, 2015, Kraft announced its merger with Heinz, arranged by Berkshire Hathaway and 3G Capital. The resulting Kraft-Heinz Company is the fifth largest food company in the world. Berkshire Hathaway became a majority owner of Heinz on June 18, 2015. After exercising a warrant to acquire 46 195 652 shares of common stock for a total price of

$461 956.52, Berkshire increased its stake to 52.5%. The companies completed the merger on July 2, 2015.

History

1) 19th-century origins

Heinz trade started from the 19th century, promoting various products. The Heinz Company was founded by and is named for Henry J. Heinz, who was born in the United States to German immigrants, Anna and Heinrich Heinz. His father was originally from Kallstadt (then in Bavaria, now part of Rhineland-Palatinate), the son of a Heinz and Charlotte Louisa Trump, a great-great-aunt of the United States president, Donald J. Trump. His mother Anna was also from Bavaria, and they met in Pittsburgh.

Henry J. Heinz began packing foodstuffs on a small scale at Sharpsburg, Pennsylvania, in 1869. There he founded Heinz Noble & Company with a friend, L. Clarence Noble, and began marketing horseradish (山葵). The first product in Heinz and Noble's new Anchor Brand (a name selected for its biblical meaning of hope) was his mother Anna Heinz's recipe for horseradish. The young Heinz manufactured it in the basement of his father's former house.

The company went bankrupt in 1875. The following year Heinz founded another company, F&J Heinz, with his brother John Heinz and a cousin, Frederick Heinz. One of this company's first products was Heinz Tomato Ketchup. The company continued to grow.

In 1888, Heinz bought out his other two partners and reorganized the company as the H. J. Heinz Company. Its slogan, "57 varieties", was introduced by Heinz in 1896. Inspired by an advertisement he saw while riding an elevated train in New York City (a shoe store boasting "21 styles"), Heinz picked the number more or less at random because he liked the sound of it, selecting "7" specifically because, as he put it, of the "psychological influence of that figure and of its enduring significance to people of all ages".

2) 20th century

In 1905, H. J. Heinz was incorporated (组成公司), and Heinz served as its first president, holding that position for the rest of his life. Under his leadership, the company pioneered processes for sanitary (清洁的) food preparation, and led a successful lobbying effort in favor of the Pure Food and Drug Act in 1906. In 1908, he established a processing plant in Leamington, Ontario, Canada for tomatoes and other products. Heinz operated it until 2014, when it was sold.

Heinz was a pioneer in both scientific and "technological innovations to solve problems like bacterial contamination". He personally worked to control the "purity of his products by managing his employees", offering hot showers and weekly manicures for the women handling food. During World War I, he worked with the Food Administration.

In 1914, Heinz Salad Cream was invented in England.

In 1930, Howard Heinz, son of Henry Heinz, helped to fight the downturn of the Great Depression by selling ready-to-serve quality soups and baby food. They became top sellers.

During World War II, Jack Heinz led the company as president and CEO to aid the United Kingdom and offset (抵消) food shortages. Its plant in Pittsburgh was converted (转变) for a time to manufacture gliders (滑翔机) for the War Department.

In the postwar years, Jack Heinz expanded the company to develop plants in several nations overseas, greatly expanding its international presence. He also acquired Ore-Ida and Starkist Tuna.

In 1959, long-time Heinz employee Frank Armour Jr. was elected president and COO of H. J. Heinz Co., succeeding H. J. Heinz II. He was the first non-family member to hold the job since the company started in 1869. He became vice chairman in 1966, and later became chairman and CEO of Heinz subsidiary, Ore-Ida Foods Inc..

In 1969, Tony O'Reilly joined the company's UK subsidiary, soon becoming its managing director. He moved to Pittsburgh in 1971 when he was promoted to Senior Vice President for the North America and Pacific region. By 1973, board members Robert Burt Gookin and Jack Heinz selected him as COO and president. He became CEO in 1979 and chairman in 1987.

3) Heinz Oven-Baked Beans newspaper ad from 1919

Between 1981 and 1991, Heinz returned 28% annually, doubling the Standard & Poor's average annual return for those years. By 2000, the consolidation of grocery store chains, the spread of retailers such as Walmart, and growth of private-label brands caused competition for shelf space, and put price pressure on the company's products. The decline was also attributed to an inadequate response to broad demographic (人口统计学的) changes in the United States, particularly the growth in population among Hispanic (西班牙的) and increased spending power of African Americans.

On April 4, 1991, former U.S. Senator Henry John Heinz III, the third-generation successor to the Heinz fortune, and six other people were killed when a Bell 412 helicopter (直升机) and a Piper Aerostar with Heinz aboard collided in mid-air above Merion Elementary School in Lower Merion Township, Pennsylvania. His fortune passed to his wife, Teresa Heinz.

In 1998, Tony O'Reilly left Heinz after issues with the company's performance. He faced challenges from corporate governance groups and pension funds including CalPERS. He was succeeded by his deputy (副手), William R. Johnson.

4) 21st century

In 2001, Heinz acquired the pasta sauce, dry bouillon (肉汤) and soup business of Borden Foods. CEO William R. Johnson stated that "they fit very well with our tomato-based expertise (专门知识或技能)".

Billionaire Nelson Peltz initiated a proxy battle during 2006, culminating in a vote to place five of Peltz's nominees on the Board. After the final vote, two of the five nominees joined the Heinz Board. The new members of the board were Nelson Peltz and Matthew Craig Walsh.

In June 2008, Heinz began an advertising campaign in the UK for their new "New York Deli Mayo" products. The advertisement featured two men kissing in a family setting, which drew 200

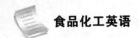

complaints to the Advertising Standards Authority. On June 24, 2008 Heinz withdrew the advertisement, which had been planned for a five-week run. The company said that some of its customers had expressed concerns. Withdrawing the advert was also controversial, with critics accusing Heinz of homophobia. The gay rights group Stonewall called for a boycott (抵制) of Heinz products. Some expressed surprise that it had responded to what they said was a relatively small number of complaints, compared to the UK's estimated 3.6 million gay and lesbian consumers. MP Diane Abbott called the decision to withdraw the advert "ill-considered" and "likely to offend the gay community".

On August 13, 2013, Heinz announced it was cutting 600 jobs in North America. On October 25, 2013, fast-food chain McDonald's announced it would end its 40-year relationship with Heinz, after the former Burger King chief Hees became its CEO.

Brands' international presence

1) The United States

The company's world headquarters are in Pittsburgh, Pennsylvania, where the company has been located since 1890. The company's "keystone" logo is based on that of Pennsylvania, the "keystone state". A majority of its ketchup is produced in Fremont, Ohio. Heinz Field was named after the Heinz Company in 2001.

Heinz opened a pickle factory in Holland, Michigan, in 1897, and it is the largest such facility in the world. The Heinz Portion Control subsidiary is located in Jacksonville, Florida, and produces single-serving containers of ketchup, mustard, salad dressings (调料), jams, jellies and syrups (糖浆).

In 2000, seven retailers, including Walmart, Albertsons, and Safeway, comprised half of the company's sales by volume.

2) Australia

Heinz Australia's head office is located in Melbourne. Products include canned baked beans in tomato sauce (popularized in the "beanz meanz Heinz" advertising campaign), spaghetti in a similar sauce, and canned soup, condensed soup, and "ready to eat" soups.

On October 6, 2008, Heinz announced plans to acquire the Australian company Golden Circle which "manufactures more than 500 products, including canned fruit and vegetables, fruit juices, drinks, cordials and jams".

On January 6, 2012, Heinz closed its tomato sauce factory in Girgarre as announced in the previous May. 146 workers lost their jobs. A local group was seeking to purchase the factory and start its own production, with offers of financial assistance from investors. The group's first offer for the site was rejected by Heinz. Girgarre was the second to last tomato sauce factory in Australia, and its closing brought an end to Heinz's 70 years of tomato processing operations in Australia.

3) Canada

Heinz was established in Canada in 1908 in a former tobacco factory in Leamington, Ontario (known as the Tomato Capital of Canada). Ketchup is the main product produced there, and the city has been a center of tomato production. The factory also produces Canada Fancy (Grade A) tomato juice, mustard, vinegar, baby food, barbecue sauces, canned pastas, beans, pasta sauces, gravies (肉汁) and soups.

On November 14, 2013, Heinz announced that the Leamington facility, the second largest in the company, would close sometime in May 2014. Ketchup processing operations were to be consolidated at the company's US locations. Over 800 local jobs were lost due to the town's largest employer ending operations there.

As a result of this corporate restructuring and the angry Canadian response, a rival food company, French's, began producing their own ketchup brand using Leamington produce. It marketed the brand with an appeal to Canadian patriotism. This successful campaign, combined with a Canadian grassroots effort on Facebook encouraging purchasing of the French's product, resulted in Heinz's market share in Canada dropping from 84% to 76%, a significant shift in a mature market.

This undesirable development was exacerbated (使恶化) in 2018 when Canadian tariffs were erected against specific American exports, which include ketchup produced in the United States, in retaliation (报复) to the US President Trump's arbitrary tariffs on Canadian steel and aluminum exports. Heinz conducted a belated (迟来的) public relations campaign in Canada to try to counter the public anger against them, a task made more difficult by public sentiment rising to encourage a boycott of American goods in reaction to US President Trump's offensive and strategic rhetoric against Canada.

4) The United Kingdom

Heinz is the leading seller of baked beans in the UK, with its beans product lines referred to as Heinz Baked Beans.

The UK headquarters moved from Hayes to the Shard in London. After opening its first overseas office in London in 1896, the company opened its first UK factory in Peckham, south London in 1905.

In 2013, the Kitt Green facility was listed as one among the world's five largest manufacturing units by the Discovery Channel (the list comprised Reliance's Jamnagar Refinery, Volkswagen's car plant, Kitt Green Foods plant, NASA's Kennedy Space Center and POSCO's steel plant). It is Europe's largest food factory and turns over more than 1 billion cans every year.

5) China

On February 22, 2013, Sanquan Food, a Chinese frozen food company, signed a contract to purchase LongFong Food, a subsidiary of Heinz Company in China. With this sale, Heinz (China) will focus on infant foods and sauces in emerging markets such as China.

(*If you want to find more information about this corporation, please log on https://en.wikipedia. org/wiki/Heinz*)

1. *Read the passage quickly by using the skills of skimming and scanning. And choose the best answer to the following questions.*

1) According to the text,_____ of Heinz ranked on the top in market in the US with a market share in excess of 50%.

    A. Ore-Ida label   B. ketchup

    C. salad dressings   D. mustard

2) On March 25, 2015, Heinz was announced to merge with _____, the fifth largest food company in the world at that year.

    A. Walmart   B. 3G Capital

    C. Kraft   D. Burger King

3) In 19th century, Henry J. Heinz founded Heinz Noble & Company with a friend, L. Clarence Noble, at _____.

    A. Sharpsburg, Pennsylvania   B. Berlin, German

    C. Bavaria   D. Rhineland-Palatinate

4) Which of the following statement is NOT true?

    A. Heinz Noble & Company kept growing until today.

    B. Heinz founded another company, F&J Heinz, in 1876.

    C. Since the foundation, F&J Heinz never went bankrupt until today.

    D. One of this company's first products was Heinz Tomato Ketchup.

5) In 1930, Howard Heinz sold ready-to-serve quality soups and baby food _____.

    A. for technological innovations

    B. to maximize its profit

    C. to edge in the market competition

    D. against the downturn of the Great Depression

6) _____ was the first non-family member to hold the job since the company started in 1869.

    A. Robert Burt Gookin   B. Frederick Heinz

    C. Frank Armour Jr.   D. Tony O'Reilly

7) By 2000, price pressure on the company's products came from the following aspects except _____.

    A. the consolidation of grocery store chains

    B. an inadequate response to broad demographic changes

    C. the interior management mistakes of the company itself

    D. growth of private-label brands

8) In 2008, the controversial advertising campaign for their new "New York Deli Mayo" products ran for _____.

    A. 200 days   B. less than a month

    C. five weeks   D. two months

9) On October 25, 2013, fast-food chain McDonald's announced it would end its 40-year relationship with Heinz after _____.

A. Heinz was facing bankruptcy

B. McDonald's found a better supplier

C. Heinz's inadequate quality hurted its fame

D. the former Burger King chief Hees became Heinz's CEO

10) Which of the following factors do not contribute to Heinz's market share loss in Canada?

A. Inadequate retailers in Canada.

B. Heinz's corporate restructuring.

C. Competition from a Canadian rival food company, French's.

D. Higher Canadian tariffs against specific American exports.

2. *In this part, the students are required to make an oral presentation on either of the following topics.*

1) Qualities to be a great entrepreneur.

2) The principles of making advertisements for a product avoiding arousing public controversy.

习题答案

I **Unit Five** **Compounds**

I. Pre-class Activity

Directions: *Please read the general introduction about Frederick Sanger and talk about his scientific belief with your classmates.*

Frederick Sanger

Frederick Sanger (13 August 1918–19 November 2013) was a British biochemist who twice won the Nobel Prize in Chemistry, one of only two people to have done so in the same category (the other is John Bardeen in Physics), the fourth person overall with two Nobel Prizes, and the third person overall with two Nobel Prizes in the sciences.

In 1958, he was awarded a Nobel Prize in Chemistry "for his work on the structure of proteins, especially that of insulin (胰岛素)". In 1980, Walter Gilbert and Sanger shared half of the chemistry prize "for their contributions concerning the determination of base sequences(基序列) in nucleic acids (核酸)". The other half was awarded to Paul Berg "for his fundamental studies of the biochemistry of nucleic acids, with particular regard to recombinant(重组)DNA".

Sanger said he found no evidence for a God so he became an agnostic(不可知论者). In an interview published in *the Times* newspaper in 2000, Sanger is quoted as saying: "My father was a committed(虔诚的) Quaker(公谊会教徒) and I was brought up as a Quaker, and for them truth is very important. I drifted away from those beliefs—one is obviously looking for truth, but one needs some evidence for it. Even if I wanted to believe in God, I would find it very difficult. I would need to see proof."

He declined the offer of a knighthood(骑士头衔), as he did not wish to be addressed as "Sir". He is quoted as saying, "A knighthood makes you different, doesn't it, and I don't want to be different." In 1986, he accepted the award of an Order of Merit(勋章), which can have only 24 living members.

Sanger died in his sleep at Addenbrookc's Hospital in Cambridge on 19 November 2013. As noted in his obituary (讣告), he had described himself as "just a chap (家伙) who messed about (瞎忙) in a lab", and "academically not brilliant".

II. Specialized Terms

Directions: *Please memorize the following specialized terms before the class so that you will be able to better cope with the coming tasks.*

anesthetic n.麻醉剂

antibacterial n.抗菌药物

artificial adj.人工的

bioavailability n.生物可利用率

biodegradable adj.可生物降解的

biomaterial n.生物材料

biotechnology n.生物技术

blood corpuscle 血细胞

canister n.(通常为金属的)小罐(装茶叶、咖啡等)

capacitor n.电容器

collagen n.胶原蛋白

comparator n.比较仪

concentration n.浓度,含量

convection n.对流

crystalline adj.透明的

current n.(水)流

cut-side n.切端,切面

defective adj.不合格的

deodorant n.除臭剂

detergent n.洗涤剂

dip v.浸

discharge v.释放

disinfectant n.消毒剂

disintegrate v.碎裂

dopant n.掺杂剂

durability n.耐久性

emulsifier n.乳化剂

endocrine adj.内分泌的

equilibrium n.平衡

extraction n.提取

feedstock n.原料

flake n.薄片

flammability n.易燃性

flavouring foods 调味食品

flexibility n.柔韧性

foaming adj.发泡的

foodstuff n.食品,粮食

gelled foodstuff 凝胶食品

gum n.牙龈

hormone n.激素

hybrid adj.混合的

hydration n.水合作用

hydrophilic adj.亲水的

hydrophobic adj.疏水的

hypoallergenic adj.低过敏的

indicator n.指示器

inflight meal 航空食品,飞机内用的便餐,机内客饭

innocuous adj.无害的

instant food 即食品

intermediate n.中间体

interplanar adj.晶面间的

isotope n.同位素

junket dessert 乳酪食品

lasting food 防腐食品,耐久存的食品

latex n.乳胶

liquefied adj.液化的

longevity n.寿命

macerate v.把……浸软

medicinal adj.药用的,治疗的

metallurgy n.冶金学

mold n.霉,霉菌

mutation n.突变,变异

nanosized adj. 纳米级的

neurotoxin n.神经毒素

neurotransmitter n.神经递质

nylon n.尼龙

objectionable constituent （食品中）有害成分

opacity n.不透光性

package n.打包

percentile n.百分位

permeability n.渗透性

pharmaceutical n.医药品

pouch food （小）袋装食品

preservative n.防腐剂

prioritize v.优先考虑

proportion n.部分;比例

puffed food 膨化食品

quantify v.量化

quantum dot 量子点

quasiparticle n.准粒子

quaternium compound 季铵盐化合物

rancid adj.（指含有油脂食物）因变质而有陈腐味道或气味的

reactor n.反应器

repellent n.拒虫剂

replicate v.复制

retractable adj.可伸缩的

sanitation n.卫生系统或设备

satchel n.纸袋,袋装食品（主要指糖果）

smoke foods 烟熏食品

snack food 休闲食品;零食;快餐食品

soda n.纯碱,苏打

soft-centered adj.软心的

spectrum n.系列

spoilage organism 腐败菌,有害微生物

susceptibility n.易感性

sustainability n.可持续性

syntheses n.合成品

synthesize v.合成

vegan adj.严守素食主义的

wholesome adj.卫生的,（食品）有益健康的

III. Watching and Listening

Task One　Molecular Formulas and Empirical Formulas

New Words

respectable adj.值得尊重的,体面的;不错的,可观的

represent v.代表;表现

formula n.公式,准则;配方奶

molecular adj.分子的

molecular formula 分子式

empirical adj.凭经验的;经验主义的;以观察或实验为依据的

empirical formula 实验式

figure out 指出,弄明白

benzene n.苯

divisor n.除数;因子

the greatest common divisor 最大公约数

structural adj.结构的;建筑的

structural formula 结构式

hexagon n.六边形

视频链接及文本

vertex n.顶点(复数 vertices)   neat adj.整洁的;简洁的;利索的

Exercises

1. *Watch the video for the first time and choose the best answers to the following questions.*

   1) Which of the following is not a formula to represent the molecules?

      A. Molecular formula.   B. Empirical formula.

      C. Structural formula.   D. Equal formula.

   2) What is the exact meaning of the word EMPIRICAL?

      A. People figure it out through an experiment or they observed it.

      B. People figure it out through careful calculation.

      C. People figure it out through logical analysis.

      D. People figure it out through considerate assumption.

   3) About the interchange between the molecular formula and the empirical formula, which
      of the following is correct?

      A. It's very easy to get the empirical formula if we know its molecular formula.

      B. It's very easy to get the molecular formula if we know its empirical formula.

      C. The empirical formula can offer less information than the molecular formula.

      D. The molecular formula is more useful than the empirical formula.

   4) Which of the following can offer the largest quantity of information?

      A. Molecular formula.   B. Empirical formula.

      C. Structural formula.   D. None of the above.

   5) According to the video clip, C_6H_6 is the _____ of benzene.

      A. molecular formula   B. empirical formula

      C. structural formula   D. none of the above

2. *Watch the video again and decide whether the following statements are true or false.*

   1) The empirical formula is essentially the actual number of atoms in that molecule.(   )

   2) When the speaker was young, he had trouble in understanding the word EMPIRICAL.(   )

   3) The greatest common divisor of $C_6H_{12}O_6$ is three.(   )

   4) The empirical formula tells you the ratio of the different items in the molecule.(   )

   5) The molecular formula really gives us the least information.(   )

3. *Watch the video for the third time and fill in the following blanks.*

   There are basically three major ways to represent a molecule— the 1)_____ formula,
   the empirical formula and the 2)_____ formula. The molecular formula is essentially the
   3)_____ number of atoms in that molecule. The empirical formula tells you what people
   have 4)_____ observed and it just tells the 5)_____ of the atoms to one another in a
   molecule without knowing in the exact molecule how many of that atom there are. And the
   way to go back, you can go from the molecular formula to the empirical formula very easily.
   You just find the 6)_____ common divisor of the number of atoms in the molecule. You

pretty much can't go back from the empirical formula to the molecular formula because you've 7) _____ information. The structural formula for benzene looks like a little 8) _____, where the 9) _____ of the hexagon are carbon atoms. In one word, the molecular formula just tells you what's in the molecule, while the empirical formula really gives you the 10) _____ information. It just tells the ratio of the different items in the molecule.

4. *Share your opinions with your partners on the following topics for discussion.*

1) Please describe to the class the exact definitions of the three different formulas, as well as the differences among them. An example will be appreciated in making such a presentation.

2) Do you think it is necessary for us students who study chemistry-related specialties in college to remember the above three formulas? Why or why not? Which of the above three is the most popularly used in our study? Please share them with your classmates.

Task Two　Theory of Acids and Bases

New Words

autoionize v.自电离
ionize v.电离;使离子化
equilibrium n.平衡;均衡;平静
bump off 吹掉
hydronium n.水合氢离子
aqueous adj.水的,水般的
solution n.溶解;溶液
hydroxide n.氢氧化物
constant adj.不变的;经常的 n.常数;常量
autoionization n.自电离
disassociate v.使分离
Celsius n.摄氏 adj. 摄氏的
exponent n.指数
minus adj.负的;减的
log base 对数基
biological adj.生物的;生物学的
Le Chatelier's principle 勒夏特列原理
over-complicate v.使过于复杂化

multiple adj.多重的,多样的
master n.专家
speller n.拼字的人

视频链接及文本

hydrochloric adj.氯化氢的,盐酸的
bold adj.黑(粗)体的
arrow n.箭,箭头
wishy-washy adj.(思想)不坚定、不清楚的;(颜色)淡而乏味的
float v.使漂浮
donate v.捐赠,捐献
strong acid 浓酸
hydrogen bromide 溴化氢
hydrogen iodide 碘化氢
nitric acid 硝酸
pop off 突然离去;匆匆而去
sulfuric acid 硫酸
perchloric acid 高氯酸

Exercises

1. *Watch the video for the first time and choose the best answers to the following questions.*

1) Which of the following may not come up in the process of water's autoionization?

A. H^+ B. OH^-

    C. H_3O^+ D. H_3O

2) The concentration of _____ determines whether a liquid is an acid or a base.

    A. H^+ B. OH^-

    C. O^{2-} D. NH_3^+

3) Which of the following is not a strong acid?

    A. HCl.   B. HBr.

    C. H_2CO_3.   D. H_2SO_4.

4) Which of the following is not a necessary condition if the pH of a liquid is equal to 7?

    A. The liquid must be H_2O.

    B. It should be at the temperature of 25℃.

    C. The water must be pure without any other substance.

    D. It must be measured at certain humidity.

5) What will the instructor go on talking about in the next section of the video?

    A. How a base comes into being.

    B. Other types of acids and their functions.

    C. The process of water's autoionization.

    D. How to distinguish acids from bases.

2. *Watch the video again and decide whether the following statements are true or false.*

1) If one of the hydrogens on H_2O bumps off and just joins the other water molecule, that water molecule will turn into OH^-. ( )

2) Water is obviously an aqueous solution. ( )

3) If the pH of a liquid is bigger than 7, it belongs to acid. ( )

4) The pH of a liquid may be minus or positive, which depends on the concentration of H^+. ( )

5) Besides halogens, other chemical elements can also form acid with H^+. ( )

3. *Watch the video for the third time and fill in the following blanks of the table.*

Chemical Symbols	Acid or Base	PH <7, >7, =7	Strong or Weak	Concentration of H^+/OH^- is >0
HCl				
NaOH				
H_2SO_4				
H_2O				
H_2CO_3				

4. *Share your opinions with your partners on the following topics for discussion.*

1) How many different categories can a liquid fall into? And what is the criterion to separate them into different groups? Please try to illustrate your opinions from different aspects, such as the loss or gain of electrons, the loss or gain of protons, and different

values of pH.

2）Suppose you were an H$^+$ and you accidentally fell into a base, how could you manage to survive? Please have a brainstorm and make up an amazing story to the class.

IV. Talking

Task One　Classical Sentences

Directions：*In this section, some popular sentences are supplied for you to read and to memorize. Then, you are required to simulate and produce your own sentences with reference to the structure.*

General Sentences

1. I'm going shopping because I need to buy some clothes.
   我想去逛逛商店,因为我需要买一些衣服。

2. Yesterday was such a beautiful day and we decided to go for a drive.
   昨天天气很好,所以我们开车出去玩了一趟。

3. What are you going to wear today?
   你今天打算穿什么?

4. I'm going to wear my blue suit. Is that all right?
   我要穿我的蓝色西装。怎么样?

5. I have some shirts to send to the laundry.
   我有一些衬衫要送到洗衣房去。

6. You ought to have that coat cleaned.
   你应该把那件外套洗一下。

7. I've got to get this shirt washed and ironed.
   我得把这件衬衫洗了熨烫一下。

8. All my suits are dirty. I don't have anything to wear.
   我所有的西装都脏了,我没有衣服穿了。

9. You'd better wear a light jacket. It's chilly today.
   你最好穿件薄夹克,今天很冷。

10. This dress doesn't fit me anymore.
    这件衣服我穿已经不合身了。

11. These shoes are worn-out. They've lasted a long time.
    这些鞋我穿好久了,已经磨破了。

12. Why don't you get dressed now? Put on your work clothes.
    为什么还不换装? 穿上你的工作服。

13. My brother came in, changed his clothes, and went out again.
    我哥哥进来了,换好衣服后又出门了。

14. I didn't notice you were wearing your new hat.
    我都没注意到你戴了新帽子。

15. If you want a towel, look in the linen closet.
    如果你想要毛巾，到放毛巾的壁橱里找。

16. My brother wants to learn how to dance.
    我弟弟想学跳舞。

17. Which would you rather do—go dancing or go to cinema?
    你想干什么？去跳舞还是去看电影？

18. I'd like to make an appointment to see Mr. Cooper.
    我想约好时间去看看库珀先生。

19. Would you like to arrange for a personal interview?
    你想安排一场个人采访吗？

20. Your appointment will be next Thursday at 10 o'clock.
    你的会面安排在下周四10点。

21. I can come any day except Thursday.
    除了星期四，我每天都能来。

22. He wants to change his appointment from Monday to Wednesday.
    他想将会面从周一改到周三。

23. She failed to call the office to cancel her appointment.
    她没有打电话让办公室取消她的预约。

24. I'm going to call the employment agency for a job.
    我要打电话给职业介绍所申请一份工作。

25. Please fill in this application form.
    请填写这张申请表。

26. Are you looking for a permanent/temporary position?
    你想应聘一个长期/临时职位吗？

27. I'm going to call a plumber to come this afternoon.
    我打算今天下午叫个管道工来。

28. I couldn't keep the appointment because I was sick.
    我病了，不能赴约了。

29. Please call before you come, otherwise we might not be home.
    请在你来之前打个电话，不然我们有可能不在家。

30. Will you please lock the door when you leave?
    你离开时把门锁上，好吗？

31. I went to see my doctor for a check-up yesterday.
    我昨天去医生那儿做检查了。

32. The doctor discovered that I'm a little overweight.
    医生发现我有些超重。

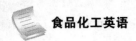

33. He gave me a chest X-ray and took my blood pressure.
    他让我做了个 X 光胸透，又给我量了量血压。

34. He told me to take these pills every four hours.
    他叮嘱我，每四小时吃一次药。

35. Do you think the patient can be cured?
    你觉得这个病人能治愈吗?

36. —What did the doctor say?
    —The doctor advised me to get plenty of exercise.
    ——医生说什么?
    ——医生建议我多做运动。

37. If I want to be healthy, I have to stop smoking.
    如果我想健康的话，我得戒烟。

38. It's just a mosquito bite. There's nothing to worry about.
    只是被蚊子叮了下，不必担心。

39. —How are you feeling today?
    —Couldn't be better.
    ——今天感觉怎样?
    ——非常好。

40. I don't feel very well this morning.
    今天早上我感到不舒服。

41. I was sick yesterday, but I'm better today.
    昨天我病了，不过今天好些了。

42. My fever is gone, but I still have a cough.
    我已经不发烧了，不过还在咳嗽。

43. —Which of your arms is sore?
    —My right arm hurts. It hurts right here.
    ——你的哪只胳膊疼?
    ——我的右胳膊疼，就在这儿。

44. —What's the matter/wrong/the trouble with you?
    —I've got a pain in my back.
    ——你哪里不舒服?
    ——我背疼。

45. —How did you break your leg?
    —I slipped on the stairs and fell down. I broke my leg.
    ——你的腿怎么伤的?
    ——我在楼梯上摔了一跤，腿就受伤了。

46. Your right hand is swollen. Does it hurt?
    你的右手肿了，疼吗?

47. It's bleeding. You'd better go to see a doctor about that cut.

你的伤口在流血,最好去医生那看看。

48. Why do you dislike the medicine so much?

为什么你这么不喜欢药?

49. I didn't like the taste of the medicine, but I took it anyway.

我不喜欢药的味道,但是无论如何我还是吃了。

50. I hope you'll be well soon.

希望你快些好。

Specialized Sentences

1. Let's say we wanted to figure out the equilibrium constant for the next reaction.

我们想知道接下来这反应的平衡常数。

2. If it wasn't dissolved and it was in the solid state, you would call this hydrogen fluoride.

如果它没溶解,而是固态,你就可以叫它氟化氢。

3. So what would the expression for the equilibrium constant look like in this situation?

那么这时平衡常数的表达式会是怎样的呢?

4. This is dependent on the concentration of the hydrofluoric acid.

这取决于氢氟酸的浓度。

5. This is indicative of the probability of this reaction happening or the probability of finding all of these molecules in the same place.

这表明发生这种反应的概率,或在同一个地方发现所有这些分子的概率。

6. These other molecules are dissolved completely in the water. So what's the solvent here?

其他分子完全溶解在水中,那么这里溶剂是什么?

7. Frankly, the concentration doesn't actually make sense for everything else.

老实说,浓度真的不是对啥都有用的。

8. So if you talk about the forward reaction, what's the forward reaction going to be dependent on?

如果是正向反应,这个正向反应取决于什么呢?

9. It's just what matters is the concentration of the water gas.

这其实只和水蒸气的浓度有关。

10. Let's say we had the reaction in which molecule A plus molecule B is in dynamic equilibrium with molecules C plus D.

假设我们有反应 A 分子和 B 分子在动态平衡中生成 C 分子和 D 分子。

11. The fact that the forward reaction rate is the same as the backward reaction rate doesn't mean that all of the concentrations are the same.

正反应速率等于逆反应速率,这并不表示所有的浓度都是一样的。

12. They're just not changing anymore because the forward and backward rates are the same.

它们只是不再改变了,因为正逆反应速率是相等的。

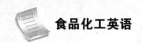

13. Given that they're at equilibrium now, what's going to happen if I add more A to the system?

这个反应处于平衡状态,那如果我向系统中加更多 A 会怎么样?

14. So more B's going to be used to consume with those extra A's you just added.

所以更多的 B 会和加进来的 A 反应而消耗掉。

15. If you took heat away from the reaction, what will be favored?

如果你减少系统的热量,反应会怎样变化?

16. Now, what's going to happen if I apply pressure to this system?

现在,如果给系统加压,会发生什么?

17. Now, when things are getting closer together, the stress of the pressure could be relieved if we end up with fewer molecules.

现在,它们都挤在一起,如果分子数减少,压强的增大就会被减弱。

18. The consistent theme we're discovering in chemistry is that everything is about random bumps between molecules.

化学探究的永恒主题是,所有东西都与分子间的随机碰撞有关。

19. When they bump randomly into each other, a lot of different things can happen in terms of knocking different parts of the molecules off each other and bonding to different things.

当它们相互随机碰撞时,能发生许多不同的变化。通过碰撞把分子的各个部分分开,然后结合成不同的物质。

20. What matters is just the ratios in these equations.

不过真正有影响的只是它们的比例。

21. We know that oxygen is a lot more electronegative than hydrogen, so it hogs all of the electrons.

氧的电负性要远远地大于氢的电负性,所以氧会占据所有的电子。

22. There's some probability that this hydrogen right here gets stuck.

这个氢有可能会被吸走。

23. Whenever hydrogen are in water, in an aqueous solution, they essentially get a ride with another water molecule.

只要氢离子是在水或其他溶液中,它们基本上就会搭上另一个水分子。

24. But the whole reason why I'm going into this whole discussion about the autoionization of water is that people really care about the concentration of—depending on how I view it—hydrogen atoms.

我展开水的自电离的整个讨论,其实是因为大家十分在意氢原子的浓度——这取决于我怎么看氢。

25. And it turns out that an acid is just something that increases the hydrogen concentration.

其实酸就是一种可以让氢的浓度升高的物质。

26. So the lower the pH, the more acid you have.

所以 pH 越低,酸性越强。

27. Remember, as long as you keep temperature constant and you're not messing too much with the molecule itself, your equilibrium constant stays constant.

记住, 只要温度不变而你又没有对分子干预过多, 平衡常数就不会变。

28. It just has one shell. But the first shell is interesting because it's complete when it only has 2 electrons.

它只有一个电子层, 但是第一电子层很有趣, 因为它只能容纳两个电子。

29. And I think you can smell some algebra coming our way.

我想你应该能够感知到接下来要谈论代数问题了。

30. I just want to make sure I'm not making a mistake by putting a minus there.

我只是想确认一下我没有因为放了个减号在这而出错。

31. So this is 0.032, which is very close to our approximation before.

这是 0.032, 非常接近我们之前的估算值。

32. I went to Wikipedia, and they have a little chart for almost any compound you look for.

我查了维基百科, 网页上有一个小图表, 几乎包含所有你要查的化合物。

33. Hydrogen without its one electron is just a proton because it has no neutrons.

没有电子的氢只是一个质子, 因为氢没有中子。

34. The fact that it disassociates completely implies that this molecule is more basic than water.

它完全水解这一事实, 说明这个分子比水碱性更强。

35. We've seen the case where A could be an NH_3, right?

我们见过 A 是氨(NH_3)的例子, 是吗?

36. This thing could grab a hydrogen from the surrounding water and become neutral then.

它可能从周围的水分子掠夺一个氢, 然后变成中性。

37. Remember, whenever I say pluck the hydrogen, just the proton, not the electron for the hydrogen.

记住, 无论什么时候我拽走氢, 只是拽走了质子, 而不包括氢的电子。

38. If we were dealing with a strong acid or a strong base, this would not be an equilibrium reaction.

如果我们正在处理强酸或者强碱, 这不会是平衡反应。

39. Even though they are the conjugate base, they wouldn't do anything.

尽管它们是共轭碱, 它们也不会产生任何影响。

40. In either reaction for given concentrations, I'm going to end up with the same concentration.

在浓度既定的两种反应中, 最终得出的浓度是一致的。

41. So let's just think a little bit about what would happen to this equilibrium if we were to stress it in some way.

那么我们就来想一下, 如果我们向某一边推动反应, 平衡会如何变化。

42. When the equilibrium moves in that direction, a lot of the hydrogenions will be replaced.

但是当平衡向那个方向移动的时候, 很多的氢离子会被取代。

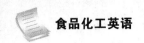

43. When you increase the amount of OH⁻, you would decrease the pH, right?

当你增加溶液中氢氧根时,你就降低了 pH,对吗?

44. Hopefully you're getting a little bit familiar with that concept right now, but if it confuses you, just play around with the logs a little bit and you'll eventually get it.

希望你现在已经比较熟悉这个概念了。如果有点儿转不过来,只要多去熟悉一下对数,你最终就会明白。

45. Even though you threw this strong base into this solution, it didn't increase the pH as much as you would have expected.

虽然你往溶液中加入了强碱,溶液 pH 值的增长量并不会像你预期那么大。

46. Everything we're dealing with right now is an aqueous solution.

我们目前要处理的所有东西都是水溶液。

47. If you were to add a strong acid to the solution, what would happen?

如果你往溶液中加入强酸会怎么样?

48. They're just going to merge with these and turn into water and become part of the aqueous solution.

它们彼此中和,转化成水,成为溶液的一部分。

49. And that's why it's called a buffer because it provides a cushion on acidity.

这就是为什么它被称为缓冲剂,因为它对溶液的酸性提供了一种缓冲。

50. It took me a couple minutes to do it, but if you just do it really fast on paper, you don't have to talk it through the way I did.

我刚才花了几分钟推导,但如果你在纸上写快些的话,你不需要像我一样一步一步来。

Task Two　Sample Dialogue

Directions: *In this section, you are going to read several times the following sample dialogue about the relevant topic. Please pay special attention to five C's (culture, context, coherence, cohesion and critique) in the dialogue and get ready for a smooth communication in the coming tasks.*

In a chemistry class

(A teacher and his students are talking about classification of substances.)

Teacher：　Good morning, boys and girls. Welcome to my class. It is commonly believed that the world is made up of abundant substances. Do you have any idea about how many categories they may fall into?

Student A：　According to purity, substances may be grouped as pure substance or mixture.

Student B：　Do you think we can find something which is 100% pure without any other elements? For example, in the gold shop, there is no gold on sale with a purity of 100%.

Student C: But, I think, with the development of technology, some pure substances free of other molecules will be made outside the laboratory.

Teacher: In the laboratory, some absolutely pure substances can be made. However, there is no existence of such pure substances. Guys, do you think impurity is necessarily bad in our work and daily life?

Student A: In some high-tech field, a high demand for purity is imposed in the process of production. If there is impurity in the fuel of rocket launching, a terrible tragedy may break out.

Student B: Yes. However, impurities sometimes make our life and work more convenient. For example, alloy. If some carbon is added into the iron, the mixed steel will be more flexible.

Student C: In order to make the natural water more delicious, some minerals have been added intentionally as impurities.

Student D: Generally speaking, impurity makes great trouble in our daily life. And it is extremely difficult for us to find in nature the pure substances without any other molecules.

Teacher: Although it is impossible to find absolutely pure substances in nature, should we deny the existence of such substances in the lab? Obviously, we cannot. Your discussion renders me thinking about the relationship between ideal and reality. Who wants to share your opinions with the class?

Student A: Terrific, I think mixture can be compared to reality while pure substances can be compared to ideal.

Student B: If we have no ideal, our life will lose its direction or goal. Adhere to the ideals once you are determined to pursue them.

Student C: Nevertheless, reality is very skinny while ideal is full. If we cannot adjust ourselves to the cold reality, our ideals will never come true.

Student D: I am sitting on the fence. On the one hand, we should stick to our ideals and don't give up easily. On the other hand, our mind should be updated regularly so that our ideals will keep in pace with the fast-developing technology.

Teacher: Today, we not only learn something about the knowledge of chemistry, specifically the classification of substances, but also share your understanding of ideal and reality. Your discussion is very fruitful and I learn quite a lot from you. Boys and girls, we should always bear in mind, live and learn. Thank you all.

Task Three   Simulation and Reproduction

Directions: *The class will be divided into three major groups, each of which will be assigned a topic. In each group, some students may be the teacher, while others may be students. In the process of discussion, please observe the principles of cooperation, politeness and choice of words. One of the*

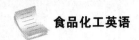

groups will be chosen to demonstrate the discussion to the class.

1）My understanding about the complicated molecular formulas.

2）My molecular formula as a person.

3）Am I acidic or alkaline?

Task Four　Discussion and Debate

Directions：*The class will be divided into two groups. Please choose your stand in regard to the following controversy and support your opinions with scientific evidences. Please refer to the specialized terms and classical sentences in the previous parts of this unit.*

In the traditional chemistry, it was universally acknowledged that something is absolutely acidic or alkaline while water is neutral. However, in the modern chemistry, it is believed that everything in the universe may be acidic or alkaline, with no existence of absolutely neutral substances. Which party do you agree with? Why?

V. After-class Exercises

1. *Dis- is a popular prefix which means "apart；lack of；to undo；to remove；not". Actually, it also popularly appears in the formation of vocabulary. Please give the Chinese meaning to the following specialized terms, as well as the verbs from which the words derive.*

Antonyms	Chinese Meaning	Original Verb	Verbs' Chinese Meaning
disassociate	失去关联	associate	（使）关联
dissolve			
disconnect			
disappear			
dislike			
disagree			
disarm			
discharge			
displace			
dismount			

2. *Fill in the following blanks with the words or phrase in the word bank. Some may be chosen more than once. Change the forms if it's necessary.*

common divisor	strong	constant	autoionize	disassociate
hexagon	molecular	concentration	exponent	multiple
arrow	ionization	empirical	structural	alkali
solution	hydronium			

1) C_6H_6 is the _____ formula of benzene, CH is _____ and the right form is the _____ formula.

2) The _____ of a liquid is the number of molecules of a substance in a given volume.

3) Nitric acid, sulfuric acid, and perchloric acid are all _____ acid.

4) If the number never changes in a chemical reaction in disregard of the temperature, the humidity or the substances involved, it is called a _____.

5) The empirical formula of benzene is a _____ with six vertices.

6) Hydrochloric is acid, while sodium hydroxide is an _____.

7) _____ is a process of converting into ions in a chemical reaction.

8) When a water molecule is combined with a hydrogen ion, it is called _____, which carries a positive charge.

9) In most liquids, when water is mixed with other chemical substances, it is called a _____.

10) If you want to go from the molecular formula to the empirical formula very easily, you just find the greatest _____ of the number of atoms in the molecule.

3. *Please fill in the following table with proper information about different formulas you've learnt in the listening section.*

Name of Substance	Molecular Formula	Empirical Formula	Structural Formula	Solid, Liquid or Gas
water	H_2O		H–O–H	
benzene				
glucose			$CH_2OH(CHOH)_4CHO$	
sulfur				

4. *Translate the following sentences into English.*

1) 正反应速率等于逆反应速率,这并不表示所有的浓度都是一样的。

_____

_____

2) 当它们相互随机碰撞时,能发生许多不同的变化。碰撞可以把分子的各个部分分开,然后结合成不同的物质。

_____

_____

3）记住,只要温度不变而你又没有对分子过多干预,平衡常数就不会变。

4）它完全水解这一事实,说明这个分子比水碱性更强。

5）虽然你往溶液中加入了强碱,溶液 pH 值的增长量并不会像你预期那么大。

5. *Please write an analysis report on the mineral water produced by Nongfu Spring Corporation. It should involve the components, as well as their percentage, the pH value and the features. Some specific data will be highly appreciated and watch out the spelling of some specialized terms you have learned in this unit.*

VI. Additional Reading

2025 Sustainability Goals of Dow Chemical Company

We have embarked on the third stage of our sustainability（可持续性）journey with our ambitious 2025 Sustainability Goals. Through these goals, we are collaborating（合作）with like-minded partners to advance the well-being of humanity by helping lead the transition to sustainable planet and society. We will lead in developing societal blueprints (蓝图) that integrate public policy solutions, science and technology, and value chain innovation to facilitate the transition to a sustainable planet and society.

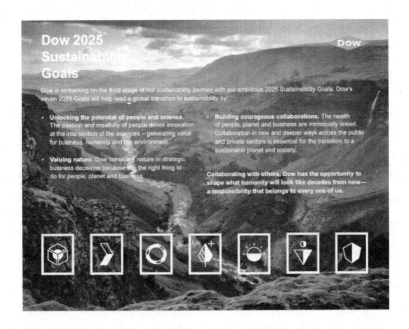

A Sustainable Future for All

Can you imagine a future where businesses, governments, and society work together to advance solutions that preserve our planet and ensure a sustainable future for all? But we know that business as usual cannot achieve this future. Our interconnected world requires a collaborative(合作的) approach to sustainability, combining meaningful actions at local and global levels. Collaborations in new and meaningful ways are the path forward to a sustainable planet.

Dow's sustainability journey has evolved from focusing on operational efficiency (Footprint), to product solutions to world challenges (Handprint), to recognizing that only through collaboration can we join others to accelerate the progress toward a sustainable planet (Blueprint).

Dow's blueprints for a sustainable planet are aligned to (对齐) the UN Sustainable Development Goals and are a collection of best practices and effective collaborations addressing the most pressing global challenges today. Currently, we are working on blueprints in water, low carbon economy, workforce development and plastic packaging that reflect our experiences and collaborations to address these challenges.

We will deliver breakthrough sustainable chemistry innovations that advance the well-being of humanity.

Dow is committed to delivering solutions essential to human progress. With more than 96

Leading the Blueprint
Dow leads in developing a societal blueprint that integrates public policy solutions, science and technology, and value chain innovation to facilitate the transition to a sustainable planet and society. Diamonds representing government, business and society unite to form a symbolic circle suspended inside a cubic system, demonstrating our planet's need for teamwork on the path to sustainability.

percent of all manufactured products enabled by chemistry, the solutions to sustainable development come down to the most basic elements in our universe. They come down to the power of chemistry. And in the hands of Dow people, our Human Element, world challenges become a universe of opportunity. As part of our 2025 Sustainability Goals, Dow will deliver breakthrough sustainable chemistry innovations that advance the well-being of humanity.

Dow delivers innovative products that make the world better. We look across the life cycle of our products to insure our innovations continue to improve, creating a world that is better because of our products. Dow's Delivering Breakthrough Innovations Goal is focused on increasing the net (净的) positive impact of our products, based upon knowledge of where our feedstocks (原料) and energy come from, our processes, how our products are used by customers, and how they are treated at end-of-life.

Delivering Breakthrough Innovations
Dow delivers breakthrough sustainable chemistry innovations that advance the well-being of humanity.
The greater-than sign signals innovation and optimism for the future; the color dynamic represents various branches of science coming together to make sustainable innovation possible.

We see a world in 2050 where healthy people, healthy communities and healthy products are the rule, not the exception. The breadth and depth of Dow's technological expertise (专业) allows us to develop game-changing (改变格局的) solutions that address energy, transportation, infrastructure (基础设施), health and nutrition, and consumer needs. Our 2025 Sustainability Goals to Deliver Breakthrough Innovations provides a new opportunity to address some of the most pressing needs of our time, by doing what Dow does best—harnessing (控制) the power of science to create solutions that positively impact the world.

We will advance a circular economy by delivering solutions to close the resource loops in key markets.

Advancing a Circular Economy
Dow advances a Circular Economy by delivering solutions to close the resource loops in key markets.
An evolution of the recycling symbol, the circle represents a continuous cycle where outputs become inputs in a system of perpetual motion and opportunity.

Many of our planet's natural resources are increasingly in short supply—fresh water is scarce, fossil fuel supplies are finite and costs for businesses, governments and society are becoming more burdensome. More than ever, it's clear we need to change our behavior and transition to a level of sustainable consumption.

Right now, we live in a primarily linear (线性的) economy where the goods we use every day are manufactured from raw materials, sold, used and then discarded as waste. Dow is leading the transition from a linear economy to one that redesigns, recycles, reuses and remanufactures to keep materials at their highest value use for as long as possible. As a result, we'll preserve our resources in a "circular (循环) economy" making the most of our natural resources. Applying the principles of a circular economy will allow us to optimize (最优化) the use and reuse of resources and ultimately reduce the amount of waste that goes into landfills (垃

圾填埋场）.

Transitioning to a circular economy is not only vital to the preservation and protection of our planet's natural resources, but also to the success of businesses at Dow. According to the Ellen MacArthur Foundation（埃伦·麦克阿瑟基金会）, circular supply chains that increase the rate of recycling, reuse and remanufacturing could generate more than $1 trillion a year by 2025. Because of our leadership position in the manufacturing of materials for use in plastic packaging and water solutions, in particular, we have a unique opportunity to take a leading role in supporting the development and implementation of the circular economy, taking into account a product's lifecycle—from creation to use to disposal—in everything we do and create.

Valuing Nature
Dow applies a business decision process that values nature, which will deliver business value and natural capital value through projects that are good for business and good for ecosystems.
The bicolored leaf exists as a sum of what is good for business and good for ecosystems. The plus sign is a nod to the potential for growth on both sides of the equation as a result of the halves coming together.

Dow applies a business-decision process that values nature, which will deliver business value and natural capital value through projects that are good for the company and better for ecosystems.

Too frequently taken for granted, nature provides a variety of valuable services—such as clean air and water—for individuals, communities and businesses. These benefits, however, are complex and can be difficult to quantify（量化）. At Dow, we're committed to making business decisions in a way that appreciates and incorporates the value of nature's services.

Increasing Confidence in Chemical Technology
Dow increases confidence in the safe use of chemical technology through transparency, dialogue, unprecedented collaboration, research and our own actions.
The colors represent human ingenuity and well-being founded on science. The sun is symbolic of Dow's newer, better, safer solutions that increase the confidence and sustainability of our products; the rays signal transparency, increasing confidence in the safe use of chemical technology.

Valuing nature can produce opportunities and drive innovation. It's a new "win-win" way of business thinking that Dow is leading today. If companies understand and value the benefits nature provides to their bottom lines, they will be more likely to plan, manage and invest in these resources in smarter, more productive and mutually beneficial ways. That's why Dow applies a business-decision process that values nature. Dow will deliver business value and natural capital value through projects that are good for the company and good for ecosystems.

Considering nature in the decision-making across a global business on the scale of the 2025 Sustainability Goals has never been done before. These goals will create new value for Dow and society by helping to sustain nature's future value. With specialized tools, Dow and other businesses around the world can incorporate the value of nature into their business processes, strategies and decisions.

We envision（想象）a future where every material we bring to market is sustainable for our people and our planet.

Chemistry helps us solve some of the world's most pressing challenges. More than 96 percent of the world's manufactured goods are enabled by chemistry, and the potential of chemistry to

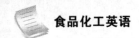

Engaging Employees for Impact
Dow people worldwide directly apply their passion and expertise to advance the well-being of people and the planet.
The juxtaposition of light and dark represents humanity and science coming together, creating a human form that represents society and the power of one to effect change.

bring social and environmental value is limitless. With this goal, we will demonstrate the value of chemistry to society, improving the ways the world understands and considers chemistry in decision-making to maximize benefits to businesses, society and the planet.

By 2025, Dow will increase confidence in the safe use of chemical technology through transparency(透明), dialogue, unprecedented (史无前例的) collaboration, research and our own actions. We want to understand and address the increasing concerns about the safe use of chemicals. We are committed to working with consumers, brand owners, retailers(零售商), researchers, regulators(调控者) and the industry to create solutions that will increase confidence and an appreciation of the important role chemical technology plays in transitioning to a sustainable planet and society. To do so, we're committed to telling the story of the positive power of chemistry. We'll also advance transparency in understanding around chemical technology, while developing safer and more sustainable products.

Multiple(多重的), simultaneous(同时的) approaches will drive us to reach this goal. We will leverage(利用) research to develop new methods to understand chemical risk and prevent it. We will bring safe products to market faster, more efficiently and with tools and data to enhance confidence in their long-term sustainability. Through regulatory advocacy(倡导) and product stewardship(管理), we will engage with governments and others to foster chemical safety through appropriate regulation and industry product stewardship. Collaborations will be formed across the value chain to promote product safety and transparency and build an understanding of the positive impact of chemical technology. We will also implement education and communication programs, internally and externally, to enhance understanding around the critical role of chemistry in our everyday lives.

Dow people worldwide directly apply their passion and expertise to advance the well-being of people and the planet.

Employee passion and talent are the force behind every philanthropic(博爱的,慈善的) engagement. Traditional and skills-based volunteerism activities demonstrate the compassion of Dow and Dow people to get engaged, and give our stakeholders(参与方,利益相关者) and communities a glimpse into the values that drive every action the company takes.

Employees apply their talents and passions to community challenges via Dow Corps, the company's overarching program for all types of employee engagement and volunteerism. Dow Corps brings together traditional and skills-based employee engagement programs to focus on priorities that are aligned to Dow's Global Citizenship efforts and the Engaging Employees for Impact goal. It is based on the premise(前提) that employees thrive when they see that their work is central to the company and the community—work that impacts both the bottom line and the greater good. Dow Corps offers two employee engagement routes:

Traditional Volunteerism—employees dedicate their time to lending a helping hand on a variety of projects ranging from building homes on a Habitat for Humanity construction project, to delivering food to people in need, to cleaning a park, to volunteering at a community event.

Skills-based Volunteerism—employees apply specific expertise, talent, professional skills and knowledge to help improve the lives of people around the world. Projects range from teaching new farming techniques to citizens in a third-world country, to working with teachers and students on specialized science activities, to implementing Dow technology to deliver safe drinking water to an

World-Leading Operations Performance
Dow maintains world-leading operations performance in natural resource efficiency, environment, health and safety.
The optical illusion created by the icon's shape and color shading indicates both a shield for safety and a building symbolic of Dow's facilities and industry.

entire community. In this realm（领域）, employees apply their expertise in the areas of business, engineering, accounting, supply chain, information technology and a host of（许多）other skills and competencies.

We will maintain world-leading operations performance in natural resource efficiency, environment, health and safety.

Dow's commitment to safety and world-leading operations performance is key to our company, our history and our "license to operate"（营业执照）in communities around the world. In 1995, we launched our 2005 Environment, Health and Safety (EH&S) Goals. With a $1 billion investment in energy efficiency, water reduction and reuse and waste reduction programs, we saved $5 billion—and fostered a culture of safety and sustainability.

The 2025 World-leading Operations Goal commits us to maintaining world-leading operations performance in natural resource efficiency, environment, health and safety. Pursuing efficient operations that drive environmental benefits for our communities and help us lead the transition to a sustainable planet and society.

Our Sustainability Journey

According to the United Nations, by the year 2050, global population will grow to nine billion people—all needing access to healthy food, clean water, sanitation（卫生）, shelter, mobility（移动）, education and healthcare（医疗）. The next few decades are pivotal（关键的）for mankind and for the planet, and the Goals were created to build a better world.

To transition toward this world, we all must change the way we live, work and play. For our part, we will do what we do best: innovate, adapt and collaborate. We must both lead by example and work with others to help lead the transition to a more sustainable planet and society.

We are proud to say that we have long been—and will remain—committed to applying our science and engineering expertise to create sustainable solutions to some of the world's greatest challenges. We are continuing to reduce our own footprint, deliver ever-increasing value to customers and society through our handprint of products and solutions, and lead in developing a blueprint for a sustainable planet and society.

The 2025 Sustainability Goals are a continuation of our ambitious 2015 Sustainability Goals,

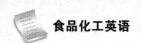

which have provided a road map for Dow since their introduction in 2006.

(*If you want to find more information about this corporation, please log on https://www.dow. com/en-us/science-and-sustainability/2025-sustainability-goals*)

1. *Read the passage quickly by using the skills of skimming and scanning. And choose the best answer to the following questions.*

   1) Which of the following is not involved in the integration of developing societal blueprints in Dow Chemical Company?

      A. Public policy solution.　　　　B. Science and technology.

      C. Value chain innovation.　　　　D. Changing the environment.

   2) What was the original focus of Dow Chemical Company in its sustainability journey?

      A. Operational efficiency.　　　　B. Product solution.

      C. World challenges.　　　　D. Employees' benefits.

   3) Which word is chosen by Dow Chemical Company to describe its 2025 Sustainability Goals?

      A. Footprint.　　　　B. Handprint.

      C. Blueprint.　　　　D. Redprint.

   4) Which of the following is Dow Chemical Company not currently working on in the blueprints?

      A. Workforce development.　　　　B. Low-carbon economy.

      C. Wlastic packaging.　　　　D. Global warming.

   5) What can Dow Chemical Company do best in the process of its development?

      A. Harness the power of science to create solutions that positively impact the world.

      B. Swallow up many small chemical corporations worldwide.

      C. Protect the global environment from being polluted severely.

      D. Produce the best chemical products in a wide range of fields.

   6) Which of the following is not true about the circular economy?

      A. Goods are manufactured from raw materials, sold, used and then discarded as waste.

      B. The circular economy can preserve our natural resources.

      C. The circular economy can redesign, recycle, reuse and remanufacture to keep materials at their highest value use for as long as possible.

      D. It can optimize the use and reuse of resources and ultimately reduce the amount of waste that goes into landfills.

   7) According to the Ellen MacArthur Foundation, how much money can be generated by 2025 through the circular supply chains?

      A. $1 trillion a year　　　　B. $1 billion a year

      C. $6 trillion a year　　　　D. $6 billion a year

   8) Why does Dow apply a business-decision process that values nature?

A. It will be punished by the government if it doesn't value nature.

B. Dow is the only company that benefits a lot from valuing nature.

C. Valuing nature can produce opportunities and drive innovation.

D. Natural resources are not inexhaustible in our universe.

9) Which of the following will not be the strategy for Dow to increase its confidence in the safe use of chemical technology?

A. Transparency.   B. Dialogue.

C. Collaboration.   D. Bargaining.

10) Which of the following is not included in the skills-based volunteerism?

A. Working with teachers and students on specialized science activities.

B. Teaching new farming techniques to citizens in a third-world country.

C. Building homes on a Habitat for Humanity construction project.

D. Implementing Dow technology to deliver safe drinking water to an entire community.

2. *In this part,the students are required to make an oral presentation on either of the following topics. When necessary, the students can log onto the Internet to search for relevant information.*

1) The history of Dow Chemical Company.

2) Focus of Dow Chemical Company's 2025 Sustainability Goals.

习题答案

Unit Six Food Safety and Health

I. Pre-class Activity

Directions: *Please read the general introduction about Louis Pasteur and tell something more about the principles of vaccination to your classmates.*

Louis Pasteur

Louis Pasteur (December 27, 1822 – September 28, 1895) was a French biologist, microbiologist and chemist renowned for his discoveries of the principles of vaccination, microbial fermentation and pasteurization. He is remembered for his remarkable breakthroughs

in the causes and prevention of diseases, and his discoveries have saved many lives ever since. He reduced mortality from puerperal fever, and created the first vaccines for rabies and anthrax. His medical discoveries provided direct support for the germ theory of disease and its application in clinical medicine. He is best known to the general public for his invention of the technique of treating milk and wine to stop bacterial contamination, a process now called pasteurization. He is regarded as one of the three main founders of bacteriology, together with Ferdinand Cohn and Robert Koch, and is popularly known as the "father of microbiology".

Pasteur was responsible for disproving the doctrine of spontaneous generation. He performed experiments that showed that without contamination, microorganisms could not develop. Under the auspices of the French Academy of Sciences, he demonstrated that in sterilized and sealed flasks nothing ever developed, and in sterilized but open flasks microorganisms could grow. Although Pasteur was not the first to propose the germ theory, his experiments indicated its correctness and convinced most of Europe that it was true. Today, he is often regarded as one of the fathers of germ theory. Pasteur made significant discoveries in chemistry, most notably on the molecular basis for the asymmetry of certain crystals and racemization. Early in his career, his investigation of tartaric acid resulted in the first resolution of what is now called optical isomers. His work led the way to the current understanding of a fundamental principle in the structure of organic compounds.

II. Specialized Terms

Directions: *Please memorize the following specialized terms before the class so that you will be able to better cope with the coming tasks.*

abdominal adj.腹部的;[鱼]有腹鳍的

ackee n.西非荔枝果

almond n.杏树;杏仁色

amanita n.伞形毒菌

amygdalin n.苦杏仁苷

antidote n.解毒药

bacterial adj.细菌的

bladder n.膀胱

bleach n.漂白剂;漂白

blood lipid profiles 血脂量

castor seed n.蓖麻

chlorophyll n.叶绿素

cleanser n. 清洁剂

coagulum n.凝结物,凝固物

coma n.[医]昏迷

convulsion n.[医]抽搐;惊厥;动乱

cramp n.痛性痉挛,抽筋

culinary adj.厨房的;烹饪的

culprit n.犯过错者,罪犯

cyanide n. [化]氰化物

decay n.腐败,腐烂

dwarfism n.矮小,侏儒症

edema n.水肿

fad food 应时食品

fetus n.胎,胎儿

flavor food 风味食品

flavor-dependent food 依赖香精的食品;与香味有关的食品

flavor-independent food 不依赖香精的食品;与香味无关的食品

food infection 食品(带菌)感染

food microbiology 食品微生物学

fruit concentrate 水果浓缩物

garnish n.(食物上的)装饰菜,配菜

gastrointestinal adj.胃肠的

geriatric food 老年食品

gourmet food 鲜美食品,珍贵食品

grayanotoxin n.木藜芦毒素,灰安毒

grocery n.食品杂货;食品杂货店

hallucinogenic adj.引起幻觉的;致幻觉药

health care food 保健食品

humidity n.湿度

hygiene n.卫生学;卫生

hypercholesterolemia n.高胆固醇血症

hypocalcemia n.低钙血症

incidence n.发生率

infant food 婴儿食品

ingredient n.(烹调的)原料

inoculated yogurt 含益生菌的酸奶

insoluble adj.不能溶解的

insoluble residue 不溶残余物

intestine n.肠

invalid food 疗效食品,病人食品

Jamaica n.牙买加

junior food 幼儿食品

junket n.凝乳食品

kasher n.犹太食品,合法的卫生食品

lactic adj.乳的

lactose n.乳糖

lactose intolerant 乳糖不耐受,对乳糖过敏

laetrile n.扁桃苷制剂；维生素 B17

low-fat food 低脂(肪)食品

low-fat spread 低脂(肪)涂抹食品

luscious adj.美味的，香甜的

macrobiotic adj.(吃)健康食品的

melamine n.三聚氰胺

moldy adj.发霉的

myristicin n.肉豆蔻醚

nutmeg n.[植]肉豆蔻(树)

elderly food 老年人食品

ovary n.[解]卵巢；[植]子房

oxalic adj.酢浆草的，采自酢浆草的

perishable adj.易腐的，易变质的 n.易腐
　　食品

pet food 宠物食品

phalloid n.鬼笔菌

pip n.果核

placenta n.胎盘，胎座

pollen n.花粉

psi＝pounds per square inch 磅/平方英寸

pulpy adj.果肉状的

punch v.用拳猛击

residue n.残渣，剩余

rhododendron n.杜鹃花

rhubarb n.大黄

ricin n.篦麻毒蛋白

slaughter n.屠宰(动物)

smack n.掌掴(声)

solanine n.茄碱

spice n.香料；调味品

spinach n.菠菜

spoilage n.食物腐败，食品变质

spore n.孢子；胚种

squeeze v.榨取，挤出

stale adj.不新鲜的

staple n.主食；主要部分

starch n.淀粉

starvation n.饥饿，绝食

stroke n.中风

testa n.外种皮，(棘皮动物等的)甲壳，(介
　　壳虫等的)介壳

tetrodotoxin n.河鲀毒素

vegetarian n.素食者

venom n.(某些蛇、蝎子等分泌的)毒液

III. Watching and Listening

Task One　Ten Poisonous Foods We Like to Eat

视频链接及文本

New Words

respiratory adj.呼吸的

famine n.饥荒

blight n.(植物)疫病；破坏性因素

neurological adj.神经学的；神经病学的

seizure n.捕捉；(疾病)突然发作

festive adj.节日的，过节似的；喜庆的；欢
　　乐的

inmate n.囚犯

nasty adj.肮脏的；下流的，令人讨厌的；恶

劣的

agitation n.搅动，搅拌；焦虑
　　不安；煽动

apprehension n.不安，忧虑，忧惧

vomit v.呕吐；大量喷出

diarrhea n.腹泻

rattlesnake n.响尾蛇

spike v.用尖物刺入；打乱某人的计划

sinest adj.最凶残的

stab v.刺;将……刺入

nauseous adj.令人作呕的;讨厌的

puffer fish n.河豚

gruesome adj.可怕的,令人毛骨悚然的

pinhead n.大头针的平头;没有头脑的人

Exercises

1. *Watch the video for the first time and choose the best answers to the following questions.*

1) If stored improperly or _____ , potatoes can develop high concentrations of the toxin solanine.

A. harvested at a rainy day   B. exposed to too much sunlight

C. cooked in a wrong way   D. contaminated

2) Kids who eat a lot of _____ have been found to forget basic skills such as tying the shoelaces.

A. raw honey   B. tuna

C. mushrooms   D. cake

3) _____ is twelve thousand times more poisonous than rattlesnake venom.

A. Castor seed   B. Papaya

C. Rhubarb   D. Durian

4) Many prisons have banned nutmeg because _____ .

A. the inmates will use it to break prison   B. it is too expensive

C. it is too difficult to restore   D. it contains a hallucinogenic

5) It's illegal for the emperor of Japan to dine on _____ due to fear for safety.

A. tuna   B. rhubarb

C. puffer fish   D. castor seeds

2. *Watch the video again and decide whether the following statements are true or false.*

1) Due to pollution in our water, tuna is packed full of mercury. (　)

2) All the almonds, growing on the sweet or bitter almond trees, pack a deadly punch. (　)

3) In 1979, a Belgian diplomat died after being stabbed with a tip of an umbrella laced with a pellet of the poison. (　)

4) When Turkey invaded Romans, the hungry soldiers lost all their senses because they ate raw honey. (　)

5) Symptoms will appear instantly after consumption of the toxic mushrooms. (　)

3. *Watch the video for the third time and fill in the following blanks.*

Number ten—Potatoes

Though it might be a worldwide 1) _____ , the potato is hiding a deadly secret. If stored improperly or 2) _____ to too much sunlight, potatoes can develop high 3) _____ of the toxin solanine. When consumed, solanine can paralyze the central 4) _____ system and lead to respiratory failure.

During the Korean War, famine struck. North Koreans had no choice but to eat 5)_____ potatoes. One community saw 6)_____ people affected by the blight, with 22 dying. But don't worry too much as it's fairly easy to spot solacing potatoes thanks to chlorophyll, which turns the skin green.

Number eight—Rhubarb

The toxic leaves of the rhubarb plant contain oxalic acid, a toxic compound used in bleach, leading to a 7)_____ sensation in the mouth and 8)_____. 9)_____ rhubarb leaves can cause breathing difficulties, 10)_____ and seizures. During World War I, British authorities recommended rhubarb leaves as a substitute for vegetables unavailable during the conflict, causing numerous cases of acute poisoning and even death.

4. *Share your opinions with your partners on the following topics for discussion.*

1) Except the ten poisonous food mentioned in this video, do you know any other food which is harmful to health and should be reduced in our daily life?

2) For many times, this video illustrates that food plays an important role in war time. Can you recall more war stories in which food's role cannot be neglected in history at home and abroad?

Task Two Organic Food and Health

New Words

synthetic adj.合成的,人造的
pesticide n.杀虫剂,农药
fungicide n.杀真菌剂
meta-analysis n.元分析
rotation n.旋转,转动;轮流,循环;[农]轮作;[天]自转
carotenoid n.类胡萝卜素
polyphenol n.多酚
LDL cholesterol 低密度脂蛋白胆固醇

immune system 免疫系统
clinically adv.冷静地,客观地 视频链接及文本
organophosphate n.& adj.有机磷酸酯(现广泛用作杀虫剂、抑燃剂等)
urine n.尿
magnitude n.巨大;重大,重要;量级
kernel n.(果实的)核,仁
listeria n.利斯特菌
bacterium n.细菌(复数为 bacteria)

Exercises

1. *Watch the video for the first time and choose the best answers to the following questions.*

1) For plants, the term "organic" basically means something grown without the use of _____.

   A. greenhouse   B. synthetic fertilizers
   C. pesticides   D. both B and C

2) If a label simply says organic, it only has to contain _____ organic ingredients.

   A. 95 percent   B. 50 percent
   C. 75 percent   D. 80 percent

3) A meta-analysis conducted in 2012 found that organic crop yields are _____ on average than conventional ones.

    A. higher   B. lower

    C. equally   D. fluctuating

4) The organic foods account for only one percent of agricultural _____.

    A. market   B. profit

    C. yields   D. acreage

5) When it comes to your own health, it's really a combination of the following elements except _____.

    A. exercise   B. diet

    C. keenness on organic food   D. genetics

2. *Watch the video again and decide whether the following statements are true or false.*

1) Organic farmers can still use pesticides and fungicides to prevent insects from destroying their crops. ( )

2) Natural pesticides are always a milder health and environmental risk than man-made ones. ( )

3) Only when a label says one hundred percent organic does it contain purely organic ingredients. ( )

4) Organic fruit and veggies are more nutritious than conventionally-grown food. ( )

5) Organic food has a higher incidence of being dangerous. ( )

3. *Watch the video for the third time and fill in the following blanks.*

Many people buy organic food hoping to feel healthier and 1) _____ have a positive impact on the 2) _____. If a 3) _____ simply says organic, it only has to 4) _____ ninety-five percent organic ingredients. The label "made with organic 5) _____" on items like bread may only contain seventy percent, while "containing organic ingredients" may only have 6) _____. Only when a label says one hundred percent organic does it contain 7) _____ organic ingredients. And for what it's worth, the term of "free range" also only 8) _____ evidence of access to the outdoors for a minimum of five minutes per day.

Most surprising is that organic food has a higher incidence of being 9) _____. The organic foods 10) _____ for only one percent of agricultural acreage or space. They account for seven percent of recalled food units in 2015.

4. *Share your opinions with your partners on the following topics for discussion.*

1) The benefits and potential harms of organic food.

2) The rational attitude towards organic food.

IV. Talking

Task One   Classical Sentences

Directions：*In this section, some popular sentences are supplied for you to read and to memorize. Then, you are required to simulate and produce your own sentences with reference to the structure.*

General Sentences

1. If it doesn't rain tomorrow, I think I'll go shopping.
   如果明天不下雨, 我想去购物。

2. There's a possibility we'll go, but it all depends on the weather.
   我们有可能去, 但要看天气怎么样。

3. Let's make a date to go shopping next Thursday.
   我们约好下周四去购物吧。

4. If I have time tomorrow, I think I'll get a haircut.
   如果我明天有空, 我想去剪头发。

5. I hope I remember to ask the barber not to cut my hair too short.
   我希望我能记得叫理发师不要把我的头发剪太短。

6. If I get my work finished in time, I'll leave for New York Monday.
   如果我能及时完成工作, 我周一就去纽约。

7. Suppose you couldn't go on the trip, how would you feel?
   设想一下, 如果你不能去旅游, 你会有什么感觉?

8. What would you say if I told you I couldn't go with you?
   如果我告诉你我不能和你一起去, 你会怎么想?

9. If I buy the car, I'll have to borrow some money.
   如果我想买那辆车的话, 我就得借些钱。

10. We may be able to help you in some way.
    我们也许能在某些方面帮助你。

11. If you were to attend the banquet, what would you wear?
    如果你要参加宴会, 你会穿什么?

12. What would you have done last night if you hadn't had to study?
    如果昨天晚上你不用学习的话, 你会做什么?

13. I would have gone on the picnic if it hadn't rained.
    要不是下雨, 我就去野炊了。

14. If you had gotten up earlier, you would have had time for breakfast.
    如果你早一点起床, 就有时间吃早饭了。

15. If I had had time, I would have called you.

我要是有时间,就给你打电话了。

16. Would he have seen you if you hadn't waved to him?

要是你没向他挥手,他还能看见你吗?

17. If he had only had enough money, he would have bought that house.

他要是有足够的钱,就会买下那房子了。

18. I wish you had called me back the next day, as I had asked you to.

可惜你没有按我的要求,在第二天给我回个电话。

19. If you hadn't slipped and fallen, you wouldn't have broken your leg.

如果你没滑跌倒,你就不会摔断了腿。

20. If I have known you want to go, I would have called you.

要是我知道你想去,我就叫你了。

21. Had I known you didn't have the key, I wouldn't have locked the door.

要是我知道你没有钥匙,我就不会锁门了。

22. She would have gone with me, but she didn't have time.

她本想和我一起去的,可是她没时间。

23. If I had asked directions, I wouldn't have got lost.

要是我问一下路,就不会迷路了。

24. Even if we could have taken the vocation, we mightn't have wanted to.

即使我们可以休假,我们也许不想去呢。

25. Everything would be alright, if you had said that.

如果你是那样说的,一切都好办了。

26. Looking back on it, I wish we hadn't given in so easily.

现在回想起来,我真希望我们没有那么轻易地让步。

27. One of these days, I'd like to take a vacation.

总有一天,我要去休假。

28. As soon as I can, I'm going to change jobs.

我要尽快换个工作。

29. There's a chance he won't be able to be home for Christmas.

他可能不能回家过圣诞节了。

30. What is it you don't like the winter weather?

你为什么不喜欢冬天的天气?

31. I don't like it when the weather gets really cold.

我不喜欢天太冷。

32. The thing I don't like about driving is all the traffic on the road.

我不喜欢开车是因为路上很拥挤。

33. He doesn't like the idea of going to bed early.

他不喜欢早睡。

34. I like to play tennis, but I'm not a very good player.

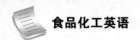
我很喜欢打网球,但是打得不是很好。

35. I don't like spinach even though I know it's good for me.
    我不喜欢菠菜,尽管我知道菠菜对我有好处。

36. I'm afraid you're being too particular about your food.
    恐怕你对食物太挑剔了。

37. He always finds fault with everything.
    他总是对每件事都吹毛求疵。

38. You have wonderful taste in clothes.
    你对衣服很有品位。

39. What's your favorite pastime?
    你最喜欢的消遣是什么?

40. What did you like best about the movie?
    你最欣赏这部电影的哪个方面?

41. The feature started at 9 o'clock and ended at 11:30.
    专题片从九点开始,一直持续到十一点半。

42. They say the new film is an adventure story.
    他们说,这部新影片讲的是一个冒险故事。

43. We went to a concert last night to hear the symphony orchestra.
    我们昨天晚上去听交响乐了。

44. A group of us went out to the theater last night.
    昨晚我们一群人去了剧院。

45. The new play was good and everybody enjoyed it.
    这个新剧很好,人人都喜欢。

46. By the time we got there, the play had already begun.
    我们到达时,戏已经开始了。

47. The usher showed us to our seats.
    引导员把我们带到了座位前。

48. The cast of the play included a famous actor.
    这场戏的演员阵容里有一位很著名的演员。

49. After the play was over, we all wanted to get something to eat.
    戏结束了以后,我们都想去吃点东西。

50. Will the new movie be welcomed by its audience or just raise another wave of dissappointment?
    那么这部新电影是会受到观众欢迎,还是会引起新一波的失望呢?

Specialized Sentences

1. A food allergy is also known as a food hypersensitivity.
   食物过敏是一种食物超敏症状。

2. A food allergy occurs when a normally harmless food elicits an immune response to the

allergen, which is usually a protein.
食物过敏是指由正常情况下无害食品引发的免疫系统对过敏原(通常为蛋白质)的反应。

3. An immune response of antibodies to fight the allergen and in turn this immune response brings upon the symptoms of food allergy.
免疫系统通过释放抗体来对抗过敏原,同时这种反应就会带来食物过敏的症状。

4. Allergic or immunological reactions are classified into IgE mediated and non-IgE mediated reactions.
这些过敏或免疫反应被分为抗体 IgE 介导型和非 IgE 介导型反应。

5. IgE is an antibody that is formed by the immune system in response to a substance that the body deemed harmful.
IgE 是免疫系统为了对抗它认为对人体有害的物质而释放的一种抗体。

6. IgE antibodies elicit an immune response causing symptoms of an allergy.
IgE 抗体会引发免疫反应。这种免疫反应会引发过敏症状。

7. If the response occurs several hours or a few days after consuming the food, it is most likely a non-IgE response.
如果在进食之后几个小时甚至是几天后才出现过敏反应,那么极有可能这是一种非 IgE 介导型反应。

8. Non-IgE response typically involves gastrointestinal symptoms such as nausea or stomach cramping.
非 IgE 介导型反应主要出现的是胃肠症状,比如恶心和胃痉挛。

9. Typically IgE mediated reactions occur a few minutes or an hour after consumption.
典型的 IgE 介导型反应一般出现在进食后几分钟到一小时的时间内。

10. Typical symptoms of IgE mediated reactions include anaphylaxis.
典型的 IgE 介导型反应症状包括过敏性反应。

11. Anaphylaxis is an acute, potentially life-threatening hypersensitivity reaction.
过敏性反应是一种严重的可能会威胁生命的过敏症状。

12. Anaphylaxis is defined by a number of signs and symptoms ranging from coughing to chest pain and the loss of consciousness.
过敏性反应一般包括以下症状:咳嗽、胸部疼痛和失去意识。

13. A food allergy can range from mild to severe.
食物过敏可能比较轻微,也可能非常严重。

14. Any food allergy may have the potential to become life-threatening.
任何一种食物过敏都会有威胁生命的可能。

15. Urticaria is a skin reaction usually accompanied by swelling and itching.
荨麻疹是一种皮肤上的反应,通常伴有肿胀和瘙痒。

16. Over the years, the safety of many food additives, from food dyes to trans fats, has come into question.

多年来,从食用色素到反式脂肪,众多食品添加剂的安全问题深受质疑。

17. Artificial food colors are chemical dyes used to color food and drinks.
人造食品色素是用来给食物和饮料上色的化学染料。

18. Many types of processed foods have artificial coloring in them.
许多类型的加工食品都含有人工色素。

19. Artificial food color is suspected of causing increased hyperactivity in children.
人们怀疑人工色素会增加儿童过度好动。

20. The dye Yellow No. 5 has been thought to worsen asthma symptoms.
黄色五号色素(柠檬黄)一直被认为会加剧哮喘症状。

21. Yellow No. 6 (sunset yellow) are approved for use in food products and must be listed as ingredients on labels.
黄色六号色素(日落黄)允许在食品中使用,并且必须在成分标签上注明。

22. High-fructose corn syrup is a sweetener made from corn, which is sweeter and cheaper than sucrose.
高果糖谷物糖浆是由玉米制成的甜味剂。它比蔗糖更甜,也更便宜。

23. Aspartame is an artificial sweetener.
天冬苯丙二肽酯是一种人工甜味剂。

24. Aspartame is a commonly used additive for sweetening diet soft drinks.
天冬苯丙二肽酯通常被用作低热量软饮料的甜味剂。

25. Large epidemiological studies haven't found a link between aspartame and cancer.
大量流行病学研究尚未发现天冬苯丙二肽酯致癌。

26. Monosodium glutamate is a form of the naturally occurring chemical glutamate, which by itself looks like salt or sugar crystals.
味精是一种天然存在的化学物质谷氨酸盐。它本身看起来像是盐或糖的结晶体。

27. Glutamate doesn't have a flavor of its own, but it enhances other flavors and imparts a savory taste.
谷氨酸盐本身没有味道,但是它能加强其他味道,并产生一种好吃的味觉。

28. Glutamate, also known as "umami", is the fifth essential flavor that the human palate can detect, in addition to sweet, salty, bitter, and sour.
谷氨酸盐,也被称之为"鲜味",是除了酸甜咸苦外,第五种人类能够品尝到的味道。

29. Tomatoes, soybeans, and seaweed are examples of foods that have a lot of glutamate naturally.
土豆、黄豆和海藻是常见的含有天然谷氨酸盐的食物。

30. Sodium benzoate is a food additive used as a preservative.
苯甲酸钠是用来防腐的一种食品添加剂。

31. Sodium nitrite is usually found in preserved meat products, like sausages and canned meats.
亚硝酸钠常见于腌肉制品中,比如香肠和肉罐头。

32. There is evidence that sodium nitrite could have been to blame for a lot of the gastric cancers.

有证据证明亚硝酸钠可能是胃癌的罪魁祸首。

33. Trans fats are created when manufacturers add hydrogen to vegetable oil.

当生产商把氢添加到蔬菜油中时,便会产生反式脂肪。

34. These "partially hydrogenated oils" are used most often for deep-frying food, and in baked goods.

这些"部分氢化"的食用油最常用于油炸和烘焙食品。

35. Trans fats are believed to increase the risk of heart disease and type 2 diabetes.

人们相信反式脂肪会增加患心脏病和2型糖尿病的风险。

36. Trans fats have been found to lower people's HDL (good) cholesterol and raise LDL (bad) cholesterol.

人们发现反式脂肪可以降低高密度脂蛋白(好)胆固醇水平,提升低密度脂蛋白(坏)胆固醇水平。

37. You can find the potassium sorbate in cheese, wine and dried meats.

你还可以在奶酪、葡萄酒和风干肉制品中发现山梨酸钾。

38. Sodium benzoate can inhibit mold and yeast from spoiling food.

苯甲酸钠可以抑制霉菌和酵母腐蚀食品。

39. Stabilizers, thickeners and texturizers give us that texture that we love when eating certain foods.

稳定剂、增稠质和调质剂能使我们吃某种食物时享受到我们喜爱的食物质地。

40. Mold, air, bacteria, fungi or yeast can easily ruin food without additives.

霉菌、空气、细菌、真菌和酵母很容易使不含添加剂的食品变质。

41. We add food additives to give food a longer shelf life and to keep the product fresher longer.

因此,添加剂的放入,能够使食品贮存时间和保鲜时间更长。

42. Food additives can also help prevent fats and oils from becoming rancid.

添加剂还能够防止油脂腐臭。

43. Things that help to improve quality would be emulsifiers, which add stability and thicken products.

提升食品质量的添加剂叫作乳化剂。它能够增加食品的稳定性、稠化食品。

44. There are two types of food additives, direct and indirect.

市场上主要有两种添加剂:直接型和间接型。

45. A direct food additive is what is added to food for a specific purpose, such as gums.

直接食品添加剂是指为了特定目的加入食品的,比如树胶。

46. By adding gum to the food it provides a better mouth feel and texture.

食品中加入树胶能够使之口感质地更精良。

47. An indirect food additive is anything that becomes part of the food unintentionally through

the packaging, storing and handling process.

间接食品添加剂是指那些在食品包装、储存、加工过程中无意间成为食品一部分的物质。

48. Certified colours are human-made to add colour and create uniformity and will not add any flavour.

合法食用色素是指人造的使食品颜色均匀,且不会改变食品味道的物质。

49. Artificial flavours are found in a number of products like flavoured juices and flavoured soda pop.

很多食品,如调味果汁和调味汽水中,都包含天然人造香料。

50. Food and Drug Administration tests, approves and monitors food additives diligently, so consumers should feel safe about the foods they eat.

食品药品监督管理局勤勉地检验、审批、监控食品添加剂,应该能让消费者对所吃食物感到安全。

Task Two   Sample Dialogue

Directions: *In this section, you are going to read several times the following dialogue which happened in an office among colleagues working overtime. Please pay special attention to five C's (culture, context, coherence, cohesion and critique) in the dialogue and get ready for a smooth communication in the coming task.*

A chat between collegues

Brian: Hey, guys, how about another cup of coffee?

Leonard: I'm in desperate need of one. I just even cannot open my eyes.

Vivian: Isn't too much caffeine technically harmful to health?

Brian: Caffeine contained in one cup of coffee may be lethal to a bug of 1g, but not 100 000 times more massive man. No need to worry.

Leonard: Yeah, there is no reported death for drinking coffee. Instead it exerts good influence on our body.

Vivian: Such as?

Leonard: Caffeine releases dopamine to make you happy and it gets rid of headaches quite positively.

Brian: Exactly. It can also increase concentration, decrease fatigue and give you better memory. This is the academic analysis.

Vivian: Sounds purely awesome.

Brian: Don't be amazed. Coffee and caffeine in it may even function as protection from cardiovascular disease, diabetes and Parkinsonism.

Leonard: Ben Franklin and Edward Lloyd loved their coffee for the same reason—it's invaluable for staying awake and concentrated when you need.

128

Vivian: Wow, coffee seems the fuel of the modern world. I'm going to grab a cup without guilt.

Brian: Why not? Go get your working smarter and faster.

Task Three   Simulation and Reproduction

Directions: *The class will be divided into three major groups, each of which will be assigned a topic and have a discussion. In the process of discussion, please observe the principles of cooperation, politeness and choice of words. One of the groups will be chosen to demonstrate the discussion to the class.*

1) Your favorite snacks.

2) A world with no food additives.

3) A significant food safety incident.

Task Four   Discussion and Debate

Directions: *The class will be divided into two groups. Please choose your stand in regard to the following controversy and support your opinions with scientific evidences. Please refer to the specialized terms and classical sentences in the previous parts of this unit.*

Humanity's relationship with what we eat is fundamental. It is believed by many that food safety is an issue of the businessmen. The more social consciousness and morality they shoulder, the safer the food will be. However, for those who insist that food safety is a government issue, it is the powerful regulation of the authorities that can guarantee the food safety. Which side are you in favor of? Support your argument with details.

V. After-class Exercises

1. *Match the English words in Column A with the Chinese meaning in Column B.*

A	B
1) hypoallergenic	a. (小)袋装食品
2) culinary	b. 浓度,含量
3) concentration	c. 厨房的;烹饪的
4) insoluble	d. 低过敏的
5) perishable	e. 不能溶解的
6) lactose	f. 卫生的,(食品)有益健康的
7) residue	g. 三聚氰胺
8) pouch	h. 乳糖
9) melamine	i. 易腐的,易变质的
10) wholesome	j. 残渣,剩余

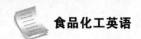

2. *Fill in the following blanks with the words or phrases in the word bank. Some may be chosen more than once. Change the forms if it's necessary.*

immune system	synthetic	hypersensitivity	asthma
heart disease	polyphenol	magnitute	type 2 diabetes
sodium benzoate	emulsifiers	immunity system	trans fats
glutamate	sodium nitrite	aspartame	

1) Trans fats are believed to increase the risk of _____ and _____.

2) Things that help to improve quality would be _____ which add stability and thicken products.

3) _____ can inhibit mold and yeast from spoiling food.

4) _____ doesn't have a flavor of its own, but it enhances other flavors and imparts a savory taste.

5) A food allergy is also known as a food _____.

6) _____ have been found to lower people's HDL (good) cholesterol and raise LDL (bad) cholesterol.

7) There is evidence that _____ could have been to blame for a lot of the gastric cancers.

8) _____ is a commonly used additive for sweetening diet soft drinks.

9) The dye Yellow No. 5 has been thought to worsen _____ symptoms.

10) IgE is an antibody that is formed by the _____ in response to a substance that the body deemed harmful.

3. *Translate the following sentences into English.*

1) 这个广泛的特异性(specificity)在酶中很有用。它在蛋白质消化中具有重要意义。

_____

_____

2) 我们需要这些营养素提供能量,构成并维持身体组织,调节生理反应。

_____

_____

3) 食品药品监督管理局(FDA)现已积极地参与到对转基因食品的安全评估中。

_____

_____

4) 世界卫生组织是一个由政府间建立并用来制订国际间食品标准的机构。

_____

_____

5) 根据生产的工艺和产品凝固的特性,酸奶可以分为两种类型:凝固型和搅拌型。

_____

_____

4 *Please write an essay of about 120 words on the topic "**How to Have a Safe and Healthy Diet**". Some specific examples will be highly appreciated and watch out the spelling of some specialized terms you have learnt in this unit.*

_____

_____

_____

_____

_____

_____

_____

_____

_____

_____

_____

_____

_____

_____

_____

VI. Additional reading

The Coca-Cola Company

The Coca-Cola Company is the world's largest beverage company. It owns or licenses and markets more than 500 non-alcoholic beverage brands, primarily sparkling beverages but also a variety of still beverages such as waters, enhanced waters, juices and juice drinks, ready-to-drink teas and coffees, and energy and sports drinks. The company owns and markets four of the world's top

five non-alcoholic sparkling beverage brands: Coca-Cola, Diet Coke, Fanta and Sprite. Finished beverage products bearing our trademarks, sold in the United States since 1886, are now sold in more than 200 countries.

The Coca-Cola Company makes branded beverage products available to consumers throughout the world through our network of company-owned or company-controlled bottling and distribution operations as well as independent bottling partners, distributors, wholesalers（批发商）and retailers（零售商）— the world's largest beverage distribution system. Beverages bearing trademarks owned by or licensed to us account for 1.9 billion of the approximately（大约）57 billion servings of all beverages consumed worldwide every day. The company believes its success depends on our ability to connect with consumers by providing them with a wide variety of options to meet their desires, needs and lifestyles.

Its success further depends on the ability of people to execute effectively, every day. The company was incorporated in September 1919 under the laws of the State of Delaware and succeeded to the business of a Georgia corporation with the same name that had been organized in 1892.

Coca-Cola Vision and Mission

The Coca-Cola Company and its bottling partners developed a 2020 Vision in 2009. This vision is a road map to doubling their global system revenues in the next 10 years by focusing on six key areas: profit, people, portfolio, partners, planet, and productivity.

Coca-Cola Goals and Objectives

The Coca-Cola Company is a leader in the beverage industry with a reputable brand and strong global presence.

According to the Coca-Cola Company's mission statement and 2020 Vision, some of its goals include:

· Increase profit by cutting down costs through productive and efficient production facilities;

· Focus on environment friendly bottling production and enforce sustainability;

· Continue to diversify its portfolio through innovations and partnerships, keeping consumer demands in mind;

· Increase annual operating income by 6%~8% in order to double their revenue（税收）by 2020.

Coca-Cola Products and Brands

"Concentrates" means flavoring ingredients and, depending on the product, sweeteners used to prepare syrups or finished beverages and includes powders for purified water products such as Dasani.

"Syrups" means the beverage ingredients produced by combining concentrates and, depending on the product, sweeteners and added water.

"Fountain syrups" means syrups that are sold to fountain（源头）retailers, such as restaurants and convenience stores, which use dispensing（分配）equipment to mix the syrups with sparkling or still water at the time of purchase to produce finished beverages that are served in cups or glasses for immediate consumption.

"Sparkling beverages" means non-alcoholic ready-to-drink beverages with carbonation（碳酸

化作用), including carbonated energy drinks and carbonated waters and flavored waters.

"Still beverages" means non-alcoholic beverages without carbonation, including noncarbonated waters, flavoured waters and enhanced waters, noncarbonated energy drinks, juices and juice drinks, ready-to-drink teas and coffees, and sports drinks.

"Company Trademark Beverages" means beverages bearing our trademarks and certain other beverage products bearing trademarks licensed to us by third parties for which we provide marketing support and from the sale of which we derive economic benefit.

"Trademark Coca-Cola Beverages" or "Trademark Coca-Cola" means beverages bearing the trademark Coca-Cola or any trademark that includes Coca-Cola or Coke (that is, Coca-Cola, Diet Coke and Coca-Cola Zero and all their variations and line extensions, including Coca-Cola Light, caffeine-free Diet Coke, Cherry Coke, etc.).

Coca-Cola SWOT Analysis

Strengths:

1. Advertising and marketing capabilities

The Coca-Cola Company's annual advertising spending was US $3.499 billion, US $3.266 billion and US $3.342 billion in 2014, 2013 and 2012 respectively. Advertising expenses accounted for 6.9% of total revenues each year. In 2014, The Coca-Cola Company was the largest advertiser in the beverage industry in the world. The company's large advertising budget provides competitive advantages such as:

· helping to introduce new products to the market

· promoting the brand

· informing consumers about the product's features

· communicating brand's message to the public

· increasing sales

In addition, the company's total marketing expenses reached US $7 billion (or 15.2% of total revenue in 2014), generating US $45.9 billion in revenue. It is one of the largest marketing budgets in the beverage industry and it is used very effectively. Only PepsiCo uses its marketing budget more effectively, spending just US $3.9 billion to generate US $66.7 billion in revenue.

2. Strength in size and financial clout

The company wields considerable financial muscle with strong acquisition capability and

funds for marketing, which are crucial for expansion and retaining consumer loyalty during a time of difficult market conditions.

3. Global strength from geographic(地理学的) spread

The company has a very strong global geographic mix. Large volumes of sales are made outside core developed markets, vastly improving the company's resilience to regional economic downturns.

Weaknesses:

1. Heavy reliance on carbonates

While the company has expanded its soft drinks portfolio of late, it remains highly dependent on carbonates. Hence, it is vulnerable to a downturn in this category.

2. Image from "high sugar" carbonates

As consumer awareness of the risks of a high sugar diet has grown, the sugar content of regular Coca-Cola could hinder the company as the consumer agenda increasingly switches to looking for healthier food and drink options.

Opportunities:

1. Health and wellness

A growing trend in soft drinks is that of health and wellness. Consumers increasingly understand the importance of healthy diets and are actively looking for healthier drinking options.

2. Emerging markets

India and China, in particular, represent good volume opportunities. Carbonates, juice drinks and bottled water should be the focus for the company.

Threats:

1. Strong competition

PepsiCo remains a considerable foe (敌人) in soft drinks and having stolen a march in terms of bottler buyouts (收购), it has possibly a crucial advantage in terms of repositioning itself in a redefined (重新定义) marketplace.

2. Cannibalisation(同类相食) in low calorie categories

In some cases, and in particular in low calorie carbonates, the company's brands can sometimes be found in direct competition with each other and at risk of cannibalisation.

Culture of Coca-Cola Company

Behavior of the company: The company's inclusive culture is defined by our seven core values—leadership, passion, integrity, collaboration, diversity, quality, and accountability. Its central promise at The Coca-Cola Company is to refresh the world in mind, body, and spirit, and inspire moments of optimism; to create value and make a difference. Two assets give us the opportunity to keep this promise—its people and its brand. The Coca-Cola Company leverages (利用) a worldwide team that is rich in diverse people, talent and ideas.

As a global business, the company's ability to understand, embrace (拥抱) and operate in a multicultural world—both in the marketplace and in the workplace—is critical to our sustainability. Its diversity workplace strategy includes programs to attract, retain (保持), and develop diverse talent; provide support systems for groups with diverse backgrounds; and educate all associates so that we master the skills to achieve sustainable growth. We work hard to ensure an inclusive and fair work environment for our associates, all of whom undergo diversity training on a regular basis. We find ongoing dialogue leads to better understanding of our colleagues, our suppliers, our customers, our stakeholders, and ultimately, to greater success in the marketplace.

Equal Opportunity Policy

The Coca-Cola Company values all employees and the contributions they make. Consistent with this value, the company reaffirms its long-standing commitment to equal opportunity and affirmative (肯定的) action in employment, which are integral parts of our corporate environment. The company strives to create an inclusive work environment free of discrimination and physical or verbal (言语的) harassment (骚扰) with respect to race, sex, colour, national origin, religion, age, disability, sexual orientation, gender identity and/or expression, genetic information or veteran status (退役军人). We will make reasonable accommodations in the employment of qualified individuals with disabilities, for religious beliefs and whenever else appropriate.

The company maintains equal employment opportunity functions to ensure adherence to all laws and regulations, and to the company policy in the areas of equal employment opportunity and affirmative action. All managers are expected to implement and enforce the company policy of non-discrimination, equal employment opportunity and affirmative action, as well as to prevent acts of harassment within their assigned (分配) area of responsibility. Further, it is part of every individual's responsibility to maintain a work environment that reflects the spirit of equal opportunity and prohibits (禁止) harassment.

Key Strategic Objectives and Challenges

1. Increasing revenue streams from all fronts

In order to achieve its goal of doubling the revenue in ten years, Coca-Cola needs to sell its products in new geographic areas and expand its product to meet the consumers' changing preference and behaviors. Maintaining its current market size in the developed market, the company also needs to increase sales in developing markets.

2. Diversification

Carbonated beverages are the company's *bread and butter* business so that the company heavily relies on their sales. This implies that the company needs to increase awareness and sales on other drinks, such as bottled water, juice, ready-to-drink tea, and even Asian specialty drinks since the consumer preferences are changing. Moreover, in order to maintain their share of sales in the increasing competitive market, Coca-Cola has to continue to strengthen their

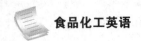

brand loyalty, innovation, and expand into other product categories in the beverage industry.

3. Diet products cannibalizing standard variants

As consumers have growing concerns about their health, such as obesity issues, which results in a reducing demand for standard cola. Therefore, the amount of sugar in regular soft drinks needs to be reduced accordingly. Although the introduction of the diet cola successfully addressed this issue, the increasing demand and sales of diet drinks cannibalized the sales of standard cola. The company needs to find a way to sustain their revenues while anticipating consumers' preference changes.

4. Acquisition targets in developed markets

With the strong penetration power in the mature soft drinks industry, the Coca-Cola Company's revenue growth can be generated from secondary markets or new markets. However, in developed markets, an acquisition option is limited because of market consolidation. It is challenging for the company to make large acquisitions in all markets.

Position of Company in Next 10 Years

The Coca-Cola Company has a strong history. Coca-Cola has 30% market share of the global beverage industry. If the company maintains its global market share up to 2020, it will add $90 billion to its market cap based on the expected increase in global beverage value. Coca-Cola has a market cap of $183 billion, giving it an expected compound growth rate of 6.9% (not including dividends and share repurchases) up to 2020 if it does not gain market share. We believe Coca-Cola will continue to gain market shareand reward shareholders with share repurchases (回购) and dividends (红利), driving up the company's CAGR into the double digits for the next several years.

Recommendations

1. Continue with expansion of low calorie drinks, particularly low calorie carbonates

Governments all over the world are publicizing obesity concerns and the sugar content of soft drinks has come under increasing scrutiny (细查). The Coca-Cola Company (TCCC) has already developed low and zero calorie drinks that are the top-ranked (排名顶尖的) brands within their categories. However, the company should continue to research low calorie drinks and in particular, natural low calorie options such as those using Stevie and more recent development in monk fruit (罗汉果).

2. Capitalize on the wider health and wellness trend within soft drinks markets

Areas such as "better for you", "naturally healthy" and "fortified (增强)/functional" fit well within soft drinks formats. TCCC already has a range of juices and waters that could prove to be future winners for the company, but it must continue to develop and innovate new HW angles within its brands. A key growth area could be fortified/functional drinks, where opportunities to add new functional ingredients are continually being discovered.

3. Spread risk from carbonates—look to push other drinks categories

The popularity of carbonated soft drinks has waned (衰落) in many markets, or has at least seen a slowdown in growth. TCCC should look to expand its sales of other beverages. Categories in which the company is not dominant globally include RTD tea. Pushing development in these categories would provide a holistic (整体的) growth opportunity into the all-important Chinese and Asia Pacific territories. Dairy is another area TCCC should explore further following the equity investment in Core Power.

Company's Scope after 10 Years

Coca-Cola is the gold standard in non-alcoholic beverage companies. Coca-Cola has significant growth potential ahead in developing markets, especially India and China. As consumer preferences slowly shift from sparkling to still beverages, Coca-Cola has positioned itself as the dominant still beverage company in the world.

Shareholders of Coca-Cola will likely be rewarded in the future with a double digit compound annual growth rate resulting from dividends, share repurchases, growing market share, and tailwinds (顺风) from growth in the overall beverage industry in the developing world. Coca-Cola often ranks as a top 10 dividend growth stock for long-term investors using The 8 Rules of Dividend Investing.

(*If you want to find more information about this corporation, please log on https://bohatala. com/coca-cola-organizational-behavior-project-report/.*)

1. *Read the passage quickly by using the skills of skimming and scanning. And choose the best answer to the following questions.*

   1) Which of the following is not the product of the Coca-Cola Company?

   A. Coffee.   B. Water.

   C. Beer.   D. Sports drinks.

   2) The company believes its success depends on _____.

   A. the ability to satisfy consumers' needs

   B. the attractive price

   C. the financial power

   D. the research on products

   3) _____ means non-alcoholic ready-to-drink beverages with carbonation.

   A. Concentrates   B. Sparkling beverages

   C. Syrups   D. Trademark Coca-Cola

   4) From 2012 to 2014, the Coca-Cola Company's annual advertising spending was never less than _____.

   A. US $3.499 billion   B. US $3.342 billion

   C. US $7 billion   D. US $3.266 billion

   5) Which of the following is one of the weaknesses of the Coco-Cola Company?

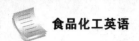

A. Heavy reliance on carbonates.   B. Geographic spread.

C. Emerging markets.   D. Strong competition.

6) As a global business, the company's ability to _____, both in the marketplace and in the workplace, is critical to our sustainability.

    A. leverage a national team

    B. refresh the world in mind, body, and spirit

    C. inspire moments of optimism

    D. understand, embrace and operate in a multicultural world

7) The Equal Opportunity Policy for managers in The Coca-Cola Company reflects in the following aspects except _____.

    A. non-discrimination

    B. same level of savages for all workers

    C. prevent acts of harassment

    D. equal employment opportunity and affirmative action

8) What's the meaning of the italicized part in "Carbonated beverages are the company's *bread and butter* business"?

    A. Breakfast.   B. A brand of Coca-Cola's product.

    C. Means of making a living.   D. Branch in other areas.

9) With the strong penetration power in the mature soft drinks industry, the Coca-Cola Company's revenue growth can be generated from _____.

    A. local markets

    B. European markets

    C. developed markets

    D. secondary markets or new markets

10) According to the text, Coca-Cola has significant growth potential ahead in developing markets, especially _____ in the next 10 years.

    A. Jamaica and Laos   B. India and China

    C. Japan and Vietnam   D. Cambodia and Indonesia

2. *In this part, you are required to make an oral presentation on either of the following topics.*

1) Your preference of Coca-Cola or PepsiCo.

2) The invention of Coco-Cola.

习题答案

Unit Seven Organic and Inorganic Matters

I. Pre-class Activity

Directions: *Please read the general introduction about Jons Jakob Berzelius and tell something more about the great scientist to your classmates.*

Jons Jakob Berzelius

Jons Jakob Berzelius, one of the most illustrious of modern chemists, was born on the 20th of August 1779, at a farm near Wafversunda, in Ostergötland, Sweden.

In 1803 he became a professor of physics, and by his lectures rapidly founded a new, rational school of physiology, and threw new light on many difficult points connected with the chemical and physical characteristics of animal life. In the same year, he published his essay on the Division of Salts through Galvanism, in which he propounds the electrochemical theory.

Berzelius was the first person to make the distinction between organic compounds (those containing carbon), and inorganic compounds. In particular, he advised Gerardus Johannes Mulder in his elemental analyses of organic compounds such as coffee, tea, and various proteins. The term protein itself was coined by Berzelius, after Mulder observed that all proteins seemed to have the same empirical formula and came to the erroneous conclusion that they might be composed of a single type of very large molecule. Berzelius proposed the name because the material seemed to be the primitive substance of animal nutrition that plants prepare for herbivores.

In 1842, while he was engaged in a chemical experiment, an explosion took place and he was much injured, but recovered and continued to work on till the close of his days. He died on August 7, 1848. After Linnaeus, he is considered to be the greatest figure in science of which Sweden can boast.

II. Specialized Terms

Directions: *Please memorize the following specialized terms before the class so that you will be able to better cope with the coming tasks.*

acidolysis n.酸解(作用)

addition n.添加,加成

adduct n.加合物

allotrope n.同素异形体

amine n.胺

analogous adj.类似的

asymmetric adj.不对称的

butane n.丁烷

catalysis n.催化(作用)

concentrate v.浓缩

configuration n.构型

conjugation n.接合

decompose v.分解

dehydrate v.脱水

derivative n.衍生物

dilation n.膨胀

displacement n.取代;置换

distillation n.蒸馏

eclipse v.使失色

elimination n.消除

emulsification n.乳化作用

equivalent adj.当量的

ethane n.乙烷

ethylene n.乙烯

evaporation n.蒸发

formaldehyde n.甲醛

functional group 官能团

gauche n.构象

harsh adj.(反应)苛刻的

heterogeneous adj.非均相的

heterolysis n.异裂

homogeneous adj.均相的

homolysis n.均裂

hydrolysis n.水解(作用)

impuritiy n.杂质

inorganic adj.无机的

labile adj.不稳定的

lipophilic adj.亲脂性的

mechanism n.机理

microwave n.微波

migration n.迁移

mild adj.(反应)温和的

neutralize v.中和

nucleophilic adj.亲核的

organic adj.有机的

oxide n.氧化物

oxidizer n.氧化剂

poisoned adj.中毒的

polyolefin n.聚烯烃

primary adj.首要的,主要的

rearrangement n.重排

refine v.提炼

removal n.除去

saponification n.皂化

scheme n.图示

secondary adj.次要的;第二的

separation n.分离

stiffen v.使变硬,使僵硬

stinky tofu 臭豆腐

stoichiometric adj.化学计量的

strategy n.策略

sublimation n.升华

substituent n.取代基

sulfate n.硫酸盐

sulfur n.硫磺

sulfur-based adj.硫基

supercritical adj.超临界

susceptible adj.易受影响的

sushi n.寿司(日本料理中的一种食品)

suspended adj.悬浮的

synthetic adj.合成的

syrup n.糖浆剂

tang n.强烈的味道或气味

tart n.果馅饼,果馅糕点

tartrazine n.柠檬黄

temper v.使(金属)回火;调和

tertiary adj.第三的

testing equipment 试验设备

thermal conductivity 导热系数,导热率

tissue n.(人、动植物的)组织

tofu flower/ tofu brain 豆花,豆腐脑

toluene n.甲苯

toughen v.(使)坚韧,(使)强硬

toxicity n.毒性

trace element 微量元素

transformation n.转化

transition state 过渡态

transparency n.透明度

triturating juicer 磨碎式榨汁机

undergo v.经历,承受

utensil n.器皿,用具

vacuum sealing 真空密封

vanilla n.香草

vat n.桶,缸

victual n.食品,食物,饮料(通常用复数)

volatile adj.易挥发的

whey n.乳清

wine tannin 葡萄酒单宁

yeast n.酵母,发酵粉

III. Watching and Listening

Task One   Naming of Organic Compounds

视频链接及文本

New Words

nomenclature n.命名法   propane n.丙烷

Exercises

1. *Watch the video for the first time and choose the best answers to the following questions.*

   1) How many valence electrons does one carbon have?

   A. One.   B. Two.

   C. Three.   D. Four.

   2) How many valence electrons does one hydrogen have?

   A. One.   B. Two.

   C. Three.   D. Four.

   3) What is the term for CH_4?

   A. Propane.   B. Methane.

   C. Ethane.   D. Butane.

   4) What is the term for C_3H_8?

A. Propane.   B. Methane.

C. Ethane.   D. Butane.

2. *Watch the video again and decide whether the following statements are true or false.*

1) Organic chemistry is dealing with chains of carbons. (　　)

2) If we draw lines to structures, each of these lines shows four electrons. (　　)

3) Elements like carbon and hydrogen can share one electron together. (　　)

4) In organic chemistry, using the line diagram is the most useful way to show organic molecules. (　　)

5) Every carbon has two bonds. (　　)

3. *Watch the video for the third time and fill in the following blanks.*

  There's another simpler way to write this. You could write it like this. Let me do it in a 1)_____ colour. You literally could write it so we have three 2)_____. So one, two, three. Now, this seems ridiculously simple. How could this thing right here give you the same 3)_____ as all of these complicated ways to draw it. Well, in chemistry, in organic chemistry, in particular, let me call it a line 4)_____ or a line angle diagram. It's the simplest way and it's actually probably the most useful way to show 5)_____ of carbon or to show organic 6)_____. Once they get started, really complicated, because it's a pain to draw all of the H's. But when you see something like this, you assume that the end 7)_____ of each line have a carbon on it. So if you see something like that, you assume that there's a carbon at that 8)_____ point.

4. *Share your opinions with your partners on the following topics for discussion.*

1) Would you like to draw a line diagram for C_2H_6?

2) Could you try to present your idea about the chains of carbon and hydrogen to your partner?

Task Two Make Matter Come Alive

New Words

cellular adj. 细胞的   Lego n. 乐高(积木玩具)

minimal adj. 最小的   kit n. 装备

aggregate v. 集合   symbolic adj. 象征的

nano n. 纳米   stuff n. 填充物

视频链接及文本

Exercises

1. *Watch the video for the first time and choose the best answers to the following questions.*

1) If you want evolution, what do you need to compete?

 A. Cell.   B. Container.

 C. Microscope.   D. Beaker.

2) In inorganic Lego kit of molecules, there is no _____ inside.

A. carbon　　　　　　　　　　B. hydrogen

      C. sodium　　　　　　　　　　D. calcium

  3) If we want sunlight in experiment, what can we do to make it happen?

      A. A box with light on.　　　　B. A container with heat.

      C. A beaker with light on.　　　D. A container with light on.

  4) _____ matter is alive.

      A. Artificial　　　　　　　　　B. Natural

      C. Evolvable　　　　　　　　　D. Mechanical

  5) What is the core of Darwinism?

      A. Less is more.

      B. While there is life, there is hope.

      C. Constant dropping wears the stone.

      D. Only the fittest survive.

2. *Watch the video again and decide whether the following statements are true or false.*

  1) Selfish gene becomes selfish matter. (　　)

  2) 50 billion years ago, there is no life on earth. (　　)

  3) The matter that can replicate is alive. (　　)

  4) It is impossible that the universe doesn't need carbon to be alive. (　　)

  5) If we can make inorganic biology, we can make matter become evolvable. (　　)

3. *Watch the video for the third time and fill in the following blanks of the table.*

    This movie here shows this competition between 1) _____. Molecules are competing for stuff. They are made of the same 2) _____ but they want their shape to win. They want their 3) _____ to persist. And that is the key. If we can somehow encourage these molecules to talk to each other and make the right shapes and 4) _____, they will start to form cells that will 5) _____ and compete. If we manage to do that, forget the molecular detail. Let's zoom out what that could mean. So we have this special theory of 6) _____ that applies only to 7) _____ biology to us. If we could get evolution into the 8) _____ world, then I propose we should have a 9) _____ theory of evolution.

4. *Share your opinions with your partners on the following topics for discussion.*

  1) What do you think is the organic matter? How about the inorganic matter?

  2) According to the ideas of the chemist in the video, how could we change inorganic matter into life?

IV. Talking

Task One　Classical Sentences

Directions: *In this section, some popular sentences are supplied for you to read and to memorize.*

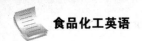

Then, you are required to simulate and produce your own sentences with reference to the structure.

General Sentences

1. Would you please tell Mr. John that I'm here?
   你能告诉约翰先生我在这吗？

2. Would you help me lift this heavy box?
   你能帮我将这个重盒子抬起来吗？

3. Please ask John to turn on the lights.
   请让约翰把灯打开。

4. Get me a hammer from the kitchen, will you?
   从厨房里给我拿个锤子，好吗？

5. Would you mind mailing this letter for me?
   你愿意帮我发这封邮件吗？

6. If you have time, will you call me tomorrow?
   如果你有时间，明天给我打电话好吗？

7. Please pick up those cups and saucers.
   请将那些杯子和碟子收拾好。

8. Will you do me a favor?
   你能帮个忙吗？

9. Excuse me, sir. Can you give me some information?
   先生，打扰一下，你能告诉我一些信息吗？

10. Do you happen to know Mr. Cooper's telephone number?
    你知道库珀先生的电话号码吗？

11. Would you mind giving me a push? My car has stalled.
    你能帮我推推车吗？我的车抛锚了。

12. Would you be so kind as to open this window for me? It's stuffy.
    你能帮我把窗户打开吗？好闷热。

13. If there's anything else I can do, please let me know.
    如果还有我能做的事情，请告诉我。

14. This is the last time I'll ever ask you to do anything for me.
    这是我最后一次麻烦你为我办事了。

15. I certainly didn't intend to cause you so much inconvenience.
    我真不想给你带来这么多不便。

16. Would you please hold the door open for me?
    帮我开门好吗？

17. You're very kind to take the trouble to help me.
    你真是太好了，不嫌麻烦来帮我。

18. I wish I could repay you somehow for your kindness.

但愿我能以某种方式报答你的好意。

19. I'm afraid it was a bother for you to do this.
做这件事恐怕会给你带来很多麻烦。

20. He'll always be indebted to you for what you've done.
对你所做的事情,他总是感激不尽。

21. Could you lend me ten dollars? I left my wallet at home.
能借我十美元吗? 我把钱包忘在家里了。

22. I'd appreciate it if you would turn off the lights. I'm sleepy.
如果你能将灯关掉的话,我会很感激。我好困。

23. You're wanted on the telephone.
有你的电话。

24. What number should I dial to get the operator?
我想接通接线员的电话,我应该拨打什么号码?

25. The telephone is ringing, would you answer it, please?
电话铃响了,你能接下电话吗?

26. Would you like to leave a message?
你想留下什么口信儿吗?

27. I have to hang up now.
我现在得挂电话了。

28. Put the receiver closer to your mouth. I can't hear you.
将话筒靠近你的嘴巴,我听不见你的声音。

29. Would you mind calling back sometime tomorrow?
你介意明天再打过来吗?

30. I almost forgot to have the phone disconnected.
我差点忘记挂断电话了。

31. It wasn't any bother. I was glad to do it.
一点儿都不麻烦,我很乐意做这件事。

32. There's just one last favor I need to ask of you.
还有最后一件事需要你的帮助。

33. I'd be happy to help you in any way I can.
很高兴我能尽我所能地帮助你。

34. Please excuse me for a little while. I want to do something.
对不起,稍等会儿,我有点事要办。

35. I didn't realize the time had passed so quickly.
我没有意识到时间过得这样快。

36. I've got a lot of things to do before I can leave.
在我走之前,我有很多事情要做。

37. For one thing, I've got to drop by the bank to get some money.

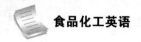

首先,我得去银行取些钱。

38. It'll take almost all my savings to buy the ticket.
买这张票,几乎花费了我所有积蓄。

39. Oh, I just remembered something! I have to apply for a passport.
我记起一件事,我得去申请个护照。

40. It's a good thing that you reminded me to take my heavy coat.
你提醒我带件厚衣服,真是太好了。

41. I would never have thought of it if you hadn't mentioned it.
要是你不提到它,我几乎想不起来了。

42. I'll see you off at the airport.
我会去机场送你。

43. Let's go out to the airport. The plane landed ten minutes ago.
我们去机场吧。飞机在十分钟前就已经着陆了。

44. There was a big crowd and we had difficulty getting a taxi.
这里人很多,我们很难打到车。

45. They're calling your flight now. You barely have time to make it.
他们现在在广播你的班机起飞时间。你勉强来得及赶上。

46. You'd better run or you're going to be left behind.
你最好跑过去,不然就会被落下了。

47. Don't forget to call us to let us know you arrived safely.
你安全到达后,别忘记打电话报个平安。

48. I'm sure I've forgotten something, but it's too late now.
我确定我忘记了什么东西,但是现在太晚了。

49. Do you have anything to declare for customs?
你有什么需要报关的吗?

50. You don't have to pay any duty on personal belongings.
你的私人物品不用交税。

Specialized Sentences

1. Organic chemistry is a chemistry subdiscipline involving the scientific study of the structure, properties, and reactions of organic compounds and organic materials.
有机化学是化学的一个分支,主要包括对有机化合物和材料的结构、性质和化学反应的科学研究。

2. Study of structure includes many physical and chemical methods to determine the chemical constitution of organic compounds and materials.
对结构的研究包括很多决定有机化合物和材料的化学构成的物理和化学方法。

3. Study of properties includes both physical properties and chemical properties.
对性质的研究包括物理和化学性质。

4. Much modern research focuses on organic chemistry involving further organometallics.

很多现代研究集中在包括有机金属化合物的有机化学。

5. Organic compounds form the basis of all earthly life and constitute a significant part of human endeavors in chemistry.

有机化合物是地球上生命的基础，也是人类在化学领域努力的重要部分。

6. Organic compounds form the basis of many commercial products.

有机化合物是很多商业产品的基础。

7. Before the nineteenth century, chemists generally believed that compounds obtained from living organisms were endowed with a vital force.

在 19 世纪以前，化学家普遍相信从生物体中获得的化合物有生命力。

8. His discovery greatly increased interest in organic chemistry.

他的发现大大增加了人们对有机化学的兴趣。

9. A crucial breakthrough for organic chemistry was the concept of chemical structure.

化学结构的概念是有机化学的重要突破。

10. It was only available to the everyday user.

它只提供给日常使用者。

11. The latter half of the 19th century witnessed systematic studies of organic compounds.

19 世纪的后半叶见证了有机化合物的系统研究。

12. The development of synthetic indigo is illustrative.

合成靛蓝染料的发展很有说明意义。

13. Converting individual petroleum compounds into different types of compounds by various chemical processes led to organic reactions.

通过不同化学过程将单一石油化合物转化成不同种化合物，带来了有机化学反应。

14. Organic compounds often exist as mixture.

有机化合物通常以混合物形式存在。

15. Organic compounds were traditionally characterized by a variety of chemical tests.

传统意义上讲，各种化学测试是有机化合物的特点。

16. Physical properties of organic compounds include both quantitative and qualitative features.

有机化合物的物理属性包括量化和质化的特点。

17. Quantitative information includes melting point, boiling point, and index of refraction.

量化信息包括熔点、沸点和折射率。

18. Qualitative properties include odor, consistency, solubility, and color.

质化属性包括气味、稠度、溶解性和颜色。

19. Organic compounds are usually not very stable at temperatures above 300 °C.

超过 300 度以上的温度时，有机化合物不是很稳定。

20. Organic compounds tend to dissolve in organic solvents.

有机化合物常常溶于有机溶剂。

21. The concept of functional group is central in organic chemistry.
官能团的概念对于有机化学很重要。

22. Functional group can have decisive influence on the chemical and physical properties of organic compounds.
官能团对有机化合物的化学和物理属性有着决定性的影响。

23. Varieties of each synthetic polymer product may exist for purposes of a specific use.
每一种合成聚合物都为了具体用途而存在。

24. Organic reactions are chemical reactions involving organic compounds.
有机反应是指包括有机化合物的化学反应。

25. Inorganic chemistry deals with the synthesis of inorganic and organometallic compounds.
无机化学处理无机和有机金属化合物的合成。

26. Physical characteristics will depend on the final composition.
物理特点取决于最终的成分。

27. This field covers all chemical compounds except the myriad organic compounds.
这个领域涵盖除有机化合物以外的所有化合物。

28. The distinction between the two disciplines is far from absolute.
两个学科之间的区别是相对的。

29. It has applications in every aspect of the chemical industry.
它在化工业的每个方面都有应用。

30. Many inorganic compounds are characterized by high melting points.
高熔点是很多无机化合物的特点。

31. Inorganic compounds are found in nature as minerals.
无机化合物在自然界中以矿物质的形式存在。

32. Traditionally, the scale of a nation's economy could be evaluated by their productivity of sulfuric acid.
传统意义上来说，可以通过硫酸的产量来衡量一个国家的经济。

33. The manufacturing of fertilizers is another practical application of industrial inorganic chemistry.
肥料的生产是另一项工业无机化学的实际应用。

34. Inorganic chemistry focuses on the classification of compounds based on their properties.
无机化学的重点是根据属性对化合物进行分类。

35. Inorganic chemistry has greatly benefited from qualitative theories.
无机化学很大程度上得益于定性理论的研究。

36. The mechanisms of reactions are discussed differently for different classes of compounds.
化合物不同，所要讨论的反应机制也就不同。

37. The mechanisms of their reactions differ from organic compounds.
它们反应的机制不同于有机化合物。

38. Inorganic chemistry is closely associated with many methods of analysis.

无机化学与很多分析方法紧密相连。

39. Although some inorganic species can be obtained in pure form from nature, most are synthesized in chemical plants and in the laboratory.

虽然有些无机物种类可以从自然界得到,但大多数是在化工厂和实验室里合成的。

40. Inorganic synthetic methods can be classified roughly according to the volatility or solubility of the component reactants.

可以根据组成反应物的挥发性或溶解性,对无机合成法进行大致分类。

41. Soluble inorganic compounds are prepared by using methods of organic synthesis.

可通过使用有机合成的方法,来准备可溶的无机化合物。

42. An inorganic compound is typically a chemical compound that lacks C-H bonds.

无机化合物实际上是缺少碳氢键的化合物。

43. There is overlap between the two fields.

两个领域有重叠。

44. Advanced interests focus on understanding the role of metals in biology and the environment.

人们对了解金属在生物学和环境中的作用越来越感兴趣。

45. These improved models led to the development of new magnetic materials and new technologies.

这些改良的模型促进新的磁性材料和新技术的发展。

46. Organic chemistry is defined as the study of carbon-containing compounds, while inorganic chemistry is the study of the remaining subset of compounds other than organic compounds.

有机化学定义为对含碳化合物的研究,而无机化学是对除了有机化合物以外的其他化合物的研究。

47. Inorganic chemists' work is based on understanding the behavior for inorganic elements and how these materials can be modified, separated and used.

无机化学家工作的基础在于理解无机元素的运动以及如何修改、分离和使用这些材料。

48. The common focus is on the exploration of the relationship between physical properties and functions.

人们普遍关注探索物理性质和功能之间的关系。

49. The raw materials needed to make most of these plastics come from petroleum and natural gas.

需要用来制作塑料的原材料,大多来自石油和天然气。

50. Because of their relatively low cost, ease of manufacture, versatility, and imperviousness to water, plastics are used in a wide range of products,

因为成本相对低廉、制作简易、用途广泛和防水,许多种产品中都会使用塑料。

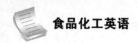

Task Two   Sample Dialogue

Directions: *In this section, you are going to read several times the following sample dialogue about the relevant topic. Please pay special attention to five C's (culture, context, coherence, cohesion and critique) in the dialogue and get ready for a smooth communication in the coming task.*

Plastics

Jack: Hi, Lily, how are you doing?

Lily: Not too bad. I'm busy with my major, a lot of compulsory and selective courses. Now we begin to learn about plastics. Do you have some ideas about it?

Jack: Plastics? Yes, plastics are widely used in everyday life.

Lily: I can't agree with you anymore. Plastics are the raw materials needed to make a range of products.

Jack: We cannot live without plastics nowadays.

Lily: That's right. Yet, do you know why plastics are so popular?

Jack: Oh, because of its low cost, I guess. And it brings great convenience to the whole market.

Lily: You get part of it. That's what we've learned in the class—the benefits of plastics.

Jack: So what are the benefits of plastics?

Lily: According to abundant researches, plastics are popular for low cost, versatility, imperviousness to water, insulation and corrosion resistance.

Jack: Wow, can you list any of its demerits?

Lily: It's time to have class. Would you like to discuss this weekend? We'll have more time to have a detailed talk.

Jack: OK, see you.

Lily: See you.

Task Three   Simulation and Reproduction

Directions: *The class will be divided into three major groups, each of which will be assigned a topic: **metals**; **glass**; **natural gas**. In the process of discussion, please observe the principles of cooperation, politeness and choice of words. Please refer to the specialized terms and classical sentences in the previous parts of this unit. One of the groups will be chosen to demonstrate the discussion to the class.*

Task Four   Speech

Directions: *Every member of the class makes a speech for one minute about your understanding of organic chemistry or inorganic chemistry. Please refer to the specialized terms and classical sentences in the previous parts of this unit.*

V. After-class Exercises

1. *Match the English words in Column A with the Chinese meaning in Column B.*

A	B
1) ethane	a. 同位素
2) functional group	b. 复制
3) replicate	c. 硫磺
4) neutralize	d. 中和
5) sublimation	e. 提炼
6) sulfur	f. 官能团
7) mechanism	g. 升华
8) isotope	h. 乙烷
9) purification	i. 机理
10) refine	j. 纯化

2. *Fill in the following blanks with the words or phrases in the word bank. Some may be chosen more than once. Change the forms if it's necessary.*

properties	compound	eclipse	oxidization	synthesis
enzyme	evaporation	refraction	concentrate	composition
aggregate	functional group	symbolic	melting point	minerals

1) Organic chemistry is a chemistry subdiscipline involving the scientific study of the structure, properties, and reactions of organic _____（化合物）and organic materials.

2) Study of structure includes many physical and chemical methods to determine the chemical _____（构成）of organic compounds and materials.

3) _____（氧化）was originally defined in terms of combination with oxygen.

4) The structure and _____（性质）of actual liquids are described.

5) Life processes are supported by the chemical reactions of complex organic compounds such as _____（酶）, hormones, proteins and lipid.

6) Quantitative information includes melting point, boiling point, and index of _____（折射）.

7) _____（官能团）can have decisive influence on the chemical and physical properties of organic compounds.

8) Inorganic chemistry deals with the _____（合成）of inorganic and organometallic compounds.

9) Many inorganic compounds are characterized by high _____（熔点）.

10) Inorganic compounds are found in nature as _____（矿物质）.

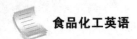

3. *Please translate the following paragraph into English.*

"有机化学"(organic chemistry)这一名词于 1806 年首次由贝采里乌斯("有机化学之父")提出。当时是作为"无机化学"的对立物而命名的。由于科学条件限制,有机化学研究的对象只能是从天然动植物有机体中提取的有机物。因而许多化学家都认为,在生物体内由于存在所谓的"生命力",才能产生有机化合物,而在实验室里是不能合成无机化合物的。

4. *Translate the following paragraph into Chinese.*

Nitrogen is a chemical element with symbol N and atomic number 7. Nitrogen has two stable isotopes: ^{14}N and ^{15}N. The first is much more common, making up 99.634% of natural nitrogen, and the second (which is slightly heavier) makes up the remaining 0.366%. This leads to an atomic weight of around 14.007 u.

5. *Please write an essay of about 100 words on the topic* "**Usage of Carbon in Our Life**". *Some specific examples will be highly appreciated and watch out the spelling of some specialized terms you have learnt in this unit.*

_____

_____

_____

_____

_____

_____

VI. Additional Reading

A Detailed Look on the Unilever

On any given (指定的) day, 2.5 billion people use Unilever products to feel good, look good and get more out of life—giving us a unique opportunity to build a brighter future.

Unilever is a British-Dutch transnational consumer goods company co-headquartered in London, the United Kingdom and Rotterdam (鹿特丹), the Netherlands. Its products include food and beverages (about 40 percent of its revenue), cleaning agents and personal care products. It is the world's largest consumer goods company measured by 2012 revenue. It is Europe's seventh most valuable company. Unilever is one of the oldest multinational companies; its products are available in around 190 countries.

Unilever owns over 400 brands, with a turnover in 2016 of over 50 billion euros, and thirteen brands with sales of over one billion euros: Axe/Lynx, Dove, Omo, Heartbrand ice creams, Hellmann's, Knorr, Lipton, Lux, Magnum, Rexona/Degree, Sunsilk and Surf. It is a dual-listed company consisting of Unilever PLC, based in London and Unilever N.V., based in Rotterdam but the company made their global headquarters in Rotterdam. The two companies operate as a single business, with a common board of directors. Unilever is organized into four main divisions—Foods, Refreshment (beverages and ice cream), Home Care, and Personal Care. It has research and development facilities (设备) in the United Kingdom, the Netherlands, China, India and the United States.

Unilever was founded in 1930 by the merger of the Dutch margarine (人造黄油) producer Margarine Unie and the British soap maker Lever Brothers. During the second half of the 20th century, the company increasingly diversified from being a maker of products made of oils and fats, and expanded its operations worldwide. It has made numerous corporate acquisitions, including Lipton (1971), Brooke Bond (1984), Chesebrough-Ponds (1987), Best Foods (2000), Ben & Jerry's (2000), Alberto-Culver (2010), Dollar Shave Club (2016) and Pukka Herbs (2017). Unilever divested its speciality chemicals businesses to ICI in 1997. In 2015, under leadership of Paul Polman, the company gradually shifted its focus towards health and beauty brands and away from food brands showing slow growth.

Unilever PLC has a primary listing on the London Stock Exchange and is a constituent of the FTSE 100 Index. Unilever N.V. has a primary listing on Euronext Amsterdam and is a constituent of the AEX Index. The company is also a lux of the Euro Stoxx 50 stock market index.

Unilever is organized into four main divisions: Personal Care (production and sale of skin care and hair care products, deodorants and oral care products); Foods [production and sale of soups, bouillons, sauces, snacks, mayonnaise (蛋黄酱), salad dressings, margarines and spreads]; Refreshment (production and sale of ice cream, tea-based beverages, weight-management products and nutritionally enhanced staples sold in developing markets); and Home Care (production and sale of home care products including powders, liquids and capsules, soap bars and other cleaning products). In the financial year ended 31 December 2013, Unilever had a total turnover (营业额) of €49.797 billion of which 36% was from Personal Care, 27% from Foods, 19% from Refreshment and 18% from Home Care. Unilever invested a total of €1.04 billion in research and development in 2013.

Unilever is one of the largest media buyers in the world, and invested around €6 billion (US $8 billion) in advertising and promotion in 2010.

Unilever's largest international competitors are Nestlé and Procter & Gamble. It also faces competition in local markets or specific product ranges from numerous companies, including Beiersdorf, ConAgra, Danone, Henkel, Mars, PepsiCo, Reckitt Benckiser and S. C. Johnson & Son. Unilever was fined by Autorité de la concurrence (一致) in France in 2016 for price-fixing on personal hygiene (卫生) products

Unilever's products include foods, beverages, cleaning agents and personal care products. The company owns more than 400 brands, which are organized into four main categories—Foods, Refreshments, Home Care, and Beauty & Personal Care. Unilever's current largest-selling brands include: Axe/Lynx, Ben&Jerry's, Dove, Heartbrand, Hellmann's/Best Foods, Knorr, Lipton, Lux/Radox, Omo/Surf, Rexona/Sure, Sunsilk, TRESemmé, Magnum, Vaseline and VO5.

Unilever has two holding companies: Unilever N.V., which has its registered and head office in Rotterdam, the Netherlands, and Unilever PLC, which has its registered office at Port Sunlight in Merseyside, the United Kingdom and its head office at Unilever House in London, the United Kingdom. Unilever PLC and Unilever N.V. and their subsidiary companies operate as nearly as practicable as a single economic entity (实体), whilst remaining separate legal entities with different shareholders and separate stock exchange listings (名单).

On 15 March 2018, Unilever announced its intention to simplify this structure by centralizing the duality of legal entities and keeping just one headquarters in Rotterdam, abandoning the London head office. Business groups and staff are unaffected, as is the dual listing.

There are a series of legal agreements between the parent companies, together with special provisions in their respective Articles of Association, which are known as the Foundation Agreements. A key requirement of the agreements is that the same people be on the Boards of the two parent companies. An Equalization Agreement regulates the mutual rights of shareholders in Unilever PLC and Unilever N.V. with the objective of ensuring that, in principle, it does not make any financial difference to hold shares in Unilever PLC rather than Unilever N.V.

In 1930, the logo of Unilever was in a sans serif (无衬线) typeface (字体) and all-caps (全部大写的). The current Unilever corporate logo was introduced in 2004 and was designed by Wolff Olins, a brand consultancy agency. The "U" shape is now made up of 25 distinct symbols, each icon representing one of the company's sub-brands or its corporate values. The brand identity was developed around the idea of "adding vitality to life".

(*If you want to find more information about this corporation, please log on https://en.wikipedia. org/wiki/Unilever*)

1. *Read the passage quickly by using the skills of skimming and scanning. And choose the best answer to the following questions.*

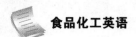

1）Which is not one of Unilever's products?

   A. Food and beverages.　　　　　　B. Cleaning agents.

   C. Personal care products.　　　　　D. Oil.

2）How many companies sell Unilever's products?

   A. 120　　　　　　　　　　　　　B. 150

   C. 190　　　　　　　　　　　　　D. 180

3）In 2015, the company gradually shifted its focus towards _____.

   A. food brands showing slow growth　　B. health and beauty brands

   C. food and beverages　　　　　　　D. financial investment

4）Which country does not have the research and development facilities?

   A. The United Kingdom.　　　　　　B. The Netherlands.

   C. Thailand.　　　　　　　　　　　D. India.

5）How much did Unilever invest in research and development in 2013?

   A. €1.04 billion.　　　　　　　　　B. €2.04 billion.

   C. €1.34 billion.　　　　　　　　　D. €3.14 billion.

6）Unilever's largest international competitors are _____.

   A. Nestlé and Procter & Gamble　　　B. Henkel

   C. Mars　　　　　　　　　　　　　D. PepsiCo

7）Unilever PLC has its registered office at Port Sunlight in Merseyside, _____ and its head office at Unilever House in London, _____.

   A. the United States; the United Kingdom

   B. the United Kingdom; the United Kingdom

   C. the United Kingdom; the United States

   D. the United States; the United States

8）In 2018, Unilever announced its intention to _____ by keeping just one headquarters in Rotterdam.

   A. change the brand's image　　　　　B. broaden its infulence

   C. simplify this structure　　　　　　D. recruit more workers

9）The Foundation Agreements requires that _____.

   A. adequate money should be spent on advertments

   B. food brands showing slow growth should be cut off as soon as possible

   C. research and development investment should be no less than €2 billion a year

   D. the same people be on the Boards of the two parent companies

10）The "U" shape is now made up of _____ distinct symbols.

   A. 21　　　　　　　　　　　　　　B. 23

   C. 25　　　　　　　　　　　　　　D. 27

2. *In this part, the students are required to make an oral presentation on either of the following topics.*

1) Your favorite Unilever's products.

2) The defects of Unilever's products.

习题答案

Unit Eight Food Marketing

I. Pre-class Activity

Directions: *Please read the general introduction about Carlos Miguel Gutierrez and have a discussion about his marketing strategy with your classmates.*

Carlos Miguel Gutierrez

Carlos Miguel Gutierrez (originally Gutiérrez; born on November 4, 1953) is an American former chairman of the Board and CEO of the Kellogg Company and once served as the 35th U.S. Secretary of Commerce from 2005 to 2009, as well as a co-chair of Albright Stonebridge Group, a strategic advisory firm.

Gutierrez joined Kellogg's in Mexico in 1975, at the age of 22, as a sales representative and management trainee. One of his early assignments was driving a delivery-truck route around local stores. Gutierrez rose through the management ranks. In January 1990, he was promoted to corporate vice president of product development at the company's headquarters in Battle Creek, Michigan, and in July of that year, he became executive vice president of Kellogg USA.

In 1999, Kellogg faced a global decline or stagnation in cereal sales. Gutierrez's strategy, known as " Volume to Value ", was to increase sales by focusing resources on higher-margin products. Higher-margin products targeted specific markets and included products such as Special K, Kashi, and Nutri-Grain bars. Extra income would fund advertising, promotions, and R&D, which would encourage further high-margin sales growth. "Volume is a means to an end—not an end," he said, "What counts is dollars."

In September 2004, *Fortune* Magazine dubbed Gutierrez as "The Man Who Fixed Kellogg", and attributed his success to "taking the slick salesmanship, financial discipline, and marketing savvy that he learned in his youth and blending it with disarming charisma, steely resolve, and an utter lack of pretension that you wouldn't expect in one so nattily dressed". The magazine also added, "He even makes golf shirts look debonair."

158

II. Specialized Terms

Directions：*Please memorize the following specialized terms before the class so that you will be able to better cope with the coming tasks.*

accumulate v.堆积,积累

adoption process 采购过程

analyzer strategy 分析者战略

apparent adj.显然的,清晰可见的

apply to 适用于;运用

appropriate adj.适当的,恰当的

as of = as from（从时间或日期）开始

assortment n.商品组合

bar code 条形码

basic item 固定畅销商品

brand loyalty 品牌忠诚度

business negotiation 商业谈判

buying inertia 购买惯性

checkout display 收银台陈列

computer-assisted ordering 计算机辅助订货

consulting services 咨询服务

continuous replenishment 持续补货

convenience store 便利店

customer retention 顾客维系

data confidentiality 数据保密

dead stock 积压库存

declining market 衰退市场

differentiated marketing 差异化营销

direct store delivery 店铺直接配送

discount store 折扣店

display n. & v.陈列

distinct adj.清楚的;种类不同的,分开的

distribution center 配送中心

distributor n.经销商,分销商

distributor system 专营分销商

engineering food 工程食品

exclusive distribution 独家分销

expansion path 扩张途径

extruded food 膨化食品

fabricated food 合成食品,组合食品

facilitate v.促进,促使

flavor industry 食品香料工业,调味料工业

food analysis 食品分析

food constituent 食品成分

food control 食品质量控制,食品质量检查

food distribution center［美］食品分销中心,批发站

food engineering 食品工程（学）

food handling 食品加工

food inspection 食品检验

food labelling 食品标志

food legislation 食品立法

food nutrition 食品营养

food packing 食品包装

food preservation 食品保藏

food product 食物产品,食品

food products regulations 食品法规

food safety 食品安全

food stamp 食品券

gondola n.货架

government regulation 政府管制

hypermarket n.特大型超级市场

inventory n.库存

inventory day 库存天数

island display 堆头式陈列

key account 重点账户

location n.位置

logo n.（公司或机构的）标识

main push item 重点销售商品

managed wholesaler 受管制批发商

merchandising n.助销

merger and acquisition 收购兼并

moderately adv.中等地;适度地

on-pack n.绑赠

out of stock 缺货

pallet display 卡板陈列

passive wholesaler 传统批发商

penetration n.渗透率

PET bottle 宝特瓶(俗称胶瓶)

price alteration 价格变更

price discount 特价

pricing n.定价

prime n.黄金时期

product life cycle 商品生命周期

promotion n.促销

purchase n.进货

quality standard 质量标准

ratio n.比率,比

running stock 储藏库存

sampling n.试吃

seasonable item 季节商品

sidekick display 侧挂陈列

slogan n.广告语

slow moving item 滞销商品

stock keeping unit 最小库存计量单位

strip display 挂条陈列

Tetra Pak 利乐无菌包装(俗称纸包装)

third party logistics 第三方物流

trademark n.(注册)商标

unit production cost 单件生产成本

universal product code 通用产品代码

vacuum-packed adj.真空包装的

value n.销售额

value share 市场份额

volume n.销售量

III. Watching and Listening

Task One　Secrets of Superbrands—Red Bull

New Words

manufacturing n.制造业

definitive adj.确定的,决定性的

hyperactive adj.极度活跃的

Thai n.泰国人,泰语

caffeine n.咖啡碱

taurine n.牛磺酸,氨基乙磺酸

repackage v.重新包装

plaster v.涂以灰泥;粘贴

race track 赛马场,跑道

hockey n.曲棍球

high-octane adj.强烈的;高辛烷值的

snoop v.探听,窥探

bomber n.轰炸机;投弹手

helicopter n.直升机

aerobatic adj.特技飞行的

endorse v.支持,核准

gigantic adj.巨大的,庞大的

sponsor v.赞助

fizzy adj.起泡发嘶嘶声的

whereas conj.然而;鉴于

property n.房地产;财产

视频链接及文本

Exercises

1. *Watch the video for the first time and choose the best answers to the following questions.*

   1) The idea of Red Bull was inspired by a drink from _____.

   A. Japan   B. Thailand

   C. Singapore   D. Cambodia

   2) According to the video clip, which of the following sports doesn't Red Bull support?

   A. Formula 1.   B. NASCAR team.

   C. Figure skating.   D. Hockey team.

   3) The video mentioned that Coke spent a gigantic amount of money on sponsoring the Olympics and _____.

   A. air race   B. basketball

   C. gymnastics   D. the World Cup

   4) Red Bull usually spends around _____ of revenue on marketing.

   A. 25 percent   B. 30 percent

   C. 10 percent   D. 35 percent

   5) To the visitor in the video, Red Bull is more a _____ than a drinks company.

   A. marketing company   B. sports company

   C. home products company   D. personal and household care

2. *Watch the video again and decide whether the following statements are true or false.*

   1) Red Bull is a brand with many manufacturing plants in Austria. (   )

   2) Mateschitz and the manufacturers split the company—seventy thirty. (   )

   3) Red Bull spends a large amount of money on high-octane sports. (   )

   4) Red Bull only discusses their marketing and promotion in sports occasionally. (   )

   5) Red Bull sponsored the Olympic Games. (   )

3. *Watch the video for the third time and fill in the following blanks.*

   For me the most 1) _____ thing about Red Bull is the amount of money they spend on getting their name plastered all over high-octane 2) _____. They own 3) _____ of Formula 1 teams, and NASCAR team, a race track, two or three football teams, hockey team and air race. The list goes on.

   I'd really like to ask Red Bull about all this 4) _____, but they told me they never discuss their 5) _____. I'm heading to their 6) _____ anyway to have a snoop around or just 7) _____ airport in Salzburg. And we hear this Hanger-7 in which Red Bull kept some 8) _____ airplanes? It was this. There's a massive hanger here. It's just got loads and loads of planes in it. I can't go in with the 9) _____, but Hanger-7 is open to 10) _____.

4. *Share your opinions with your partners on the following topics for discussion.*

   1) Have you ever drunk Red Bull? What's the taste and effect on you? How do you

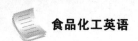

like it?

2) According to the video clip, can you summarize some particular marketing strategies of Red Bull?

Task Two　　Secrets of Superbrands—Starbucks

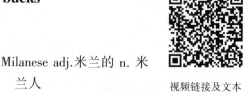
视频链接及文本

New Words

photography n.摄影,摄影术
lousy adj.讨厌的;不清洁的
flourishing adj.繁荣的;茂盛的
crossroads n.十字路口;转折点
pedestrian n.步行者,行人
expansion n.扩张
operator n.经营者;操作员
takeover n.收购
high street 主干道;高街(商业街)
Vancouver 温哥华
smash v.打碎
spark v.触发;鼓舞
mandate n.授权;命令
authentic adj.真正的;可靠的

Milanese adj.米兰的 n. 米
　兰人
humility n.谦逊,谦恭
embarrassing adj.使人尴尬的
velvet n.天鹅绒
elaborate adj.精心制作的
inspire v.激励;启迪
nurture v.养育;培育
substantial adj.大量的
quid n.一英镑,一镑金币
brainchild n.脑力劳动的产物;独创的观念
environmentalist n.环境保护论者
materialistic adj.唯物论的,实际的

Exercises

1. *Watch the video for the first time and choose the best answers to the following questions.*

1) Howard Schultz had a degree in _____.
　　A. engineering　　　　　　　B. marketing
　　C. accounting　　　　　　　D. psychology

2) With the money borrowed or raised, _____ is the most important element when they open a new shop.
　　A. size　　　　　　　　　　B. style
　　C. location　　　　　　　　D. cost

3) According to the video clip, Starbucks has around _____ Starbucks in fifty countries.
　　A. 17 000　　　　　　　　　B. 70 000
　　C. 15 000　　　　　　　　　D. 35 000

4) In order to check whether Starbucks has the original vision of what Howard Schultz announced it should be, the interviewer goes to _____.
　　A. Sydney　　　　　　　　　B. Milan
　　C. London　　　　　　　　　D. Paris

5) Starbucks charge three pounds for a cup of coffee for _____.

    A. the size of the cup   B. the styling of the cup

    C. the environment of drinking   D. all of the above

2. *Watch the video again and decide whether the following statements are true or false.*

    1) Howard Schultz was a photocopier salesman before moving on to sell coffee for a tiny company in Seattle called Starbucks. (   )

    2) A trip to London inspired Howard Schultz to buy Starbucks and turn it into a café. (   )

    3) Starbucks is seen as the McDonalds' for the middle classes. (   )

    4) The Italians totally understand what a grand double shot mocha frappuchino is. (   )

    5) Starbucks mission statement says they are to inspire and nurture the human spirit. (   )

3. *Watch the video for the third time and fill in the following blanks of the table.*

    Howard Schultz had a 1)_____ in marketing. He was a killer photocopier 2)_____ before moving on to sell coffee for a tiny company in Seattle called Starbucks. They didn't sell drinks at the time, only the dry 3)_____. Coffee in the US was out of 4)_____ and lousy. On a trip to Milan, Schultz had a coffee vision—selling the 5)_____ coffee experience to America. He went back to Seattle, borrowed a load of money from the 6)_____, bought out the owners of Starbucks, and turn the shop into a café.

    "It was the first 7)_____ of that store design, and I was very 8)_____ in that because I wanted to make sure we got it right. And the 9)_____ really needed to come to life. The romance of the beverage was going to play a more important and more 10)_____ role."

4. *Share your opinions with your partners on the following topics for discussion.*

    1) Your impression of Starbucks.

    2) Your reason to buy or not to buy a cup of Starbucks coffee.

IV. Talking

Task One   Classical Sentences

Directions: *In this section, some popular sentences are supplied for you to read and to memorize. Then, you are required to simulate and produce your own sentences with reference to the structure.*

General Sentences

1. Children enter school at the age of five, don't they?
   孩子们到五岁时就上学,是吗?

2. In elementary school, children learn to read and write.
   在小学,孩子们学习读和写。

3. In secondary school, children get more advanced knowledge.
   在中学,孩子们学到更多先进的知识。

4. In universities, students are trained to become teachers and engineers.

   在大学里,学生们被培养成老师和工程师。

5. He went to grade school in New York and high school in Chicago.

   他在纽约上小学,在芝加哥上中学。

6. In college I majored in science. What was your major?

   大学里我的专业是科学,你呢?

7. My sister graduated from high school. Graduation was last night.

   我姐姐中学毕业了。昨晚举行了毕业晚会。

8. I'm a graduate of Yale University. I have a Bachelor of Arts degree.

   我是一名耶鲁大学的毕业生。我获得了艺术学士学位。

9. If you expect to enter the university, you should apply now.

   如果你想上大学,你现在就应该申请。

10. This is my first year of college. I'm a freshman.

    这是我大学的第一年,我是新生。

11. My uncle is a high school principal.

    我叔叔是一名中学校长。

12. What kind of grades did you make in college?

    你在大学中成绩怎么样?

13. During your first year of college, did you make straight As?

    你大学一年级时,成绩全优吗?

14. My brother is a member of the faculty. He teaches economics.

    我的哥哥是一名老师。他教经济学。

15. John has many extracurricular activities. He's on the football team.

    约翰有许多课外活动。他是一名足球队员。

16. I'm a federal employee. I work for the Department of Labor.

    我是一名联邦雇员。我在劳工部工作。

17. What kind of work do you do? Are you a salesman?

    你做什么工作? 是不是销售员?

18. As soon as I complete my training, I'm going to be a bank teller.

    一旦我完成了培训,我将成为一名银行出纳员。

19. John has built up his own business. He owns a hotel.

    约翰已经有了自己的生意。他有一家旅馆。

20. What do you want to be when you grow up?

    你长大后想做什么?

21. My son wants to be a policeman when he grows up.

    我儿子长大后想当一名警察。

22. I like painting, but I wouldn't want it to be my life's work.

    我喜欢绘画,但是我不会以绘画作为终生的职业。

23. Have you ever thought about a career in the medical profession?

你是否考虑过成为一名医药行业的从业人员？

24. My uncle was a pilot with the airlines. He has just retired.

我的叔叔是航空公司的飞行员。他刚刚退休。

25. My brother's in the army. He was just promoted to the rank of major.

我的哥哥在军队。他刚刚被提升为少校。

26. I have a good-paying job with excellent hours.

我有一份工资很高、工作时间理想的工作。

27. My sister worked as a secretary before she got married.

我的姐姐结婚前是个秘书。

28. George's father is an attorney. He has his own company.

乔治的父亲是个律师，拥有自己的公司。

29. He always takes pride in his work. He's very efficient.

他总是以他的工作为荣。他是个很能干的人。

30. Mr. Smith is a politician. He's running for election as governor.

史密斯先生是个政治家，他正在为竞选州长而奔忙。

31. After a successful career in business, he was appointed ambassador.

他在生意中有了成就后，就被任命为大使。

32. Why is Mr. Smith so tired? Do you have any idea?

为什么史密斯先生这么累，你知道吗？

33. According to Mr. Green, this is a complicated problem.

听格林先生说，这是一个复杂的问题。

34. I wish you would give me a more detailed description of your trip.

我希望你能更详细地描述一下你的旅行。

35. We used to have a lot of fun when we were that age.

我们年龄那么大时经常玩得很开心。

36. I never realized that someday I would be living in New York.

我从没有想过有一天我能住在纽约。

37. We never imagined that John would become a doctor.

我们从来没有想过约翰会成为一名医生。

38. I beg your pardon. Is this seat taken?

请问，这个座位有人坐吗？

39. The waiter seems to be in a hurry to take our order.

服务员似乎急着要我们点东西。

40. —Which would you rather have, steak or fish?

　　—I want my steak well-done.

　　——你想来点什么，牛排还是鱼？

　　——我想要全熟的牛排。

41. —What kinds of vegetables do you have?

—I'll have mashed potatoes.

——你想要什么蔬菜?

——我想要些土豆泥。

42. —What do you want?

—I want a cup of coffee.

——你想要点什么?

——我想来杯咖啡。

43. Which one would you like—this one or that one?

你想要哪一个? 这一个还是那一个?

44. It doesn't matter to me.

都可以。

45. Would you please pass the salt?

麻烦把盐递过来好吗?

46. They serve good food in this restaurant.

这家餐馆的东西很好吃。

47. Are you ready for your dessert now?

现在可以吃点心了吗?

48. This knife/fork/spoon is dirty. Would you bring me a clean one, please?

这把刀/叉/勺脏了,麻烦你拿一个干净的来好吗?

49. You have your choice of three flavors of ice cream. We have vanilla, chocolate, and strawberry.

你有三种口味的冰淇淋可选:香草味的、巧克力味的和草莓味的。

50. The restaurant was filled, so we decided to go elsewhere.

这家餐馆已经没餐位了,我们得去别处了。

Specialized Sentences

1. The traditional modes of advertising are outdoor advertising, radio advertising, window display, newspaper and magazine ads, television commercials, exhibitions, etc.

传统广告模式包括户外广告、广播广告、橱窗展示、报纸和杂志广告、电视广告、展览等。

2. Newspapers and magazines are the earliest forms of press advertising.

报纸和杂志是最早的广告形式。

3. Print and broadcasting mostly come in the category of traditional forms of advertising.

印刷品和广播主要属于传统广告形式。

4. The new age advertising includes Google ads, YouTube ads, e-mail marketing, call to action ads, etc.

新时代广告包括 Google 广告、YouTube 广告、电子邮件营销、行动号召广告等。

5. Mobile and online marketing form the new age marketing tactics.

   移动和在线营销形成了新时代的营销策略。

6. Advertising is the concept to get the mass excited about product/service and compel them to purchase it.

   广告是让大众对产品/服务感到兴奋并迫使他们购买的概念。

7. Direct mailing list is prepared to send circular letters, folders, calendars, booklets and catalogues to the customers.

   直接邮寄名单被用来发送宣传邮件、文件夹、日历、小册子和目录给客户。

8. Direct Mail contains detailed information about the product, which should be attractive and convincing.

   直接邮寄广告包含有关产品的详细信息，这些信息应该具有吸引力和说服力。

9. Through this process of e-mail marketing, you can reach out to your potential customers and keep them updated about your present activities.

   通过这个电子邮件营销流程，您可以联系潜在客户并让他们了解您当前的营销活动。

10. From YouTube to Facebook to Google, every social and digital platform displays advertisements.

    从 YouTube 到 Facebook 再到 Google，每个社交和数字平台都会显示广告。

11. Before selecting a newspaper, you need to consider various factors such as coverage of the newspaper, its customer base, the cost of advertising etc.

    在选择报纸之前，您需要考虑各种因素，如报纸的覆盖范围、客户群、广告费用等。

12. Magazines may be released weekly, monthly, or annually, or might follow another form of time management.

    杂志可以每周、每月或每年发布，也可以采用其他形式的时间管理。

13. Magazine ads are glossier, descriptive, and leave a lasting impression on the readers' mind.

    杂志广告更具说服力，更具描述性，给读者留下深刻的印象。

14. Radio advertising can be explained as word-of-mouth promotion at a superior extent.

    广播广告可以解释为在更远的范围内进行口口宣传。

15. Word-of-mouth advertising involves hiring people to talk about your product or service in a public place in a way that other people overhear them.

    口口相传的广告涉及雇用人们在公共场所谈论您的产品或服务，这样旁人可以偶然听到。

16. Television commercials have the advantages of sound and sight.

    电视广告具有可视听的优点。

17. A powerful TV ad with a good script can leave an everlasting impression on the mind of the audiences.

    一个拥有好文案的强大的电视广告可以给观众留下长久的印象。

18. Pictorial presentation is more effective to connect with the customers.

图示可以更有效地与客户建立联系。

19. Television commercials is a costly medium no doubt but its mass appeal is high and so is the reach.

电视广告毫无疑问是一种代价高昂的媒体,但它的大众吸引力很高,影响范围也很广。

20. Outdoor advertising includes different media like posters, hoardings, banners, bus, train and tram advertising, etc.

户外广告包括不同的媒体,如海报,巨幅广告牌,横幅,公共汽车、火车和有轨电车广告等。

21. Outdoor advertising catches the attention of a passerby and compels them to read it.

户外广告引起路人的注意,并迫使人们阅读。

22. If your product features a notable innovation, or has an unusual feature, write a press release.

如果您的产品具有显著的创新或具有不寻常的特色,那就撰写新闻稿。

23. Canvassing the neighborhood, placing flyers in mailboxes and hanging ads on doorknobs, are good methods to target a specific area.

对邻居进行拉票,将传单放在邮箱中或在门把手上悬挂广告,是针对特定区域投放广告的好方法。

24. If the product you sell relates to your own expertise, public speaking can be a great advertisement.

如果您销售的产品与您自己的专业知识有关,公开演讲可能是一个很好的广告。

25. Importing is to bring in goods and services into a country from another country for the purpose of selling them.

进口是指将货物和服务从另一个国家引入一个国家以出售它们。

26. Exporting refers to the direct sale and marketing of goods that are domestically produced in a country to another country.

出口是指将在一国境内生产的货物直接销售给另一个国家。

27. Licensing basically allows an organization in a target country to use the property of the licensor.

许可经营基本上允许目标国家的组织使用许可人的财产。

28. This property of the licensor is intangible, such as production techniques, trademarks, and patents.

许可人的这种财产是无形的,例如生产技术、商标和专利。

29. The license pays a particular amount in exchange for the rights to utilize the intangible goods and maybe for technical assistance.

许可证支付特定金额以换取使用无形商品的权利或技术援助。

30. Franchising is the pattern of the right to use an organization's business model and brand

for a particular period of time.

特许经营是在特定时期内使用组织的商业模式和品牌的权利模式。

31. There are three historical phases of food marketing: the fragmentation phase (before 1870-1880), the unification phase (1880-1950), and the segmentation phase (1950 and later).

食品营销有三个历史阶段:分裂阶段(1870—1880 之前),统一阶段(1880-1950)和分割阶段(1950 年及以后)。

32. In the fragmentation phase, transporting food was expensive, leaving most production, distribution, and selling locally based.

在分裂阶段,运输食品很昂贵,大部分生产、分销和销售都是以当地为基础的。

33. In the fragmentation phase, the United States was divided into numerous geographic fragments.

在分裂阶段,美国被分成许多地理上的区块。

34. In the unification phase, distribution was made possible by railroads.

在统一阶段,铁路使分销成为可能。

35. In the unification phase, coordination of sales forces was made possible by telegraph and telephone.

在统一阶段,电报和电话使协调销售队伍成为可能。

36. In the unification phase, product consistency was made possible by advances in manufacturing.

在统一阶段,制造业的进步使产品品质一致成为可能。

37. In the segmentation phase, radio, television and internet advertising made the product competition more fierce.

在分割阶段,广播、电视和互联网广告使产品竞争更加激烈。

38. Distribution via the new national road system strengthened national brands.

利用新的国家道路系统进行的分销加强了国家品牌。

39. The four components of food marketing are often called the "four Ps" of the marketing mix.

食品营销的四个组成部分通常被称为营销组合的"四个 P"。

40. The four components of food marketing relate to product, price, promotion, and place.

食品营销的四个组成部分涉及产品、价格、促销和地点。

41. The product of the marketing mix refers to the goods and/or services that the organization will offer to the consumer.

营销组合的产品是指组织将向消费者提供的商品和/或服务。

42. Price encompasses the amount of money paid by the consumer in order to purchase the food product.

价格包括消费者为购买食品而支付的金额。

43. Promoting a food to consumers is done out of store, in store, and on package.

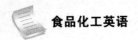

向顾客进行食品促销是在仓库外、仓库中和包装上进行的。

44. Place refers to the distribution and warehousing efforts necessary to move a food from the manufacturer to a location where a consumer can buy it.

地点是指将食品从制造商转移到消费者可以购买的地方所需的分配和仓储工作。

45. The money that manufacturers invest in the "four Ps" helps differentiate a food product.

制造商投资"四个 P"的资金有助于区分食品。

46. Advertising to children at a young age is a well-established food marketing technique designed to encourage brand preference.

向年幼儿童做广告是一种成熟的食品营销手段,旨在鼓励品牌偏好。

47. Advertising to children holds many ethical dilemmas as well.

向年幼儿童做广告也存在许多道德困境。

48. Children and teenagers have become more susceptible to unhealthy food marketing commercial messages from food organizations.

儿童和青少年更容易受到来自食品组织的不健康食品营销广告的影响。

49. Food marketing is one of the leading contributors to an increase in childhood obesity.

食品营销是导致儿童肥胖增加的主要原因之一。

50. Food marketing not only involves the marketing of products to consumers, but the reasons why consumers purchase these items and the factors influencing such choices.

食品营销不仅涉及向消费者销售产品,还涉及消费者购买这些产品的原因以及影响这些选择的因素。

Task Two Sample Dialogue

Directions: *In this section, you are going to read several times the following sample dialogue about the relevant topic. A manger is interviewing two candidates about their internship. Please pay special attention to five C's (culture, context, coherence, cohesion and critique) in the dialogue and get ready for a smooth communication in the coming task.*

In an group interview

Manager:　Good morning, guys. Both of your resumes mentioned that you have been a member of the Youth Program. What is that?

Candidate A: To be specific, it's an internship program for college students, which lasted for 5 months.

Candidate B: During that program, we went to a food company, working as a salesman in the summer vacation last year.

Manager:　What difficulties did you have during the internship?

Candidate A: During the process, our sales team was not given enough money to carry out our marketing project.

Candidate B: And we had to figure out solutions by ourselves with our teammates.

Manager： How did you feel at that time?

Candidate B： Yeah, we were frustrated at first but we pulled ourselves together soon.

Candidate A： That experience really showed us what is quite important, in order to carry out a successful marketing strategy.

Manager： So, did you find any solution?

Candidate A： Yes. We knew that our marketing project was money-consuming.

Candidate B： We revised the project and planned the budget carefully.

Manager： If you are given another chance to design and carry out a successful marketing strategy for foods, what will you take into consideration?

Candidate A： Budget definitely goes first.

Manager： That was quite a lesson, right?

Candidate A： Yeah. Except that, price and targeted customer will never be neglected.

Candidate B： The promotion scheme also needs to be carefully studied.

Task Three   Simulation and Reproduction

Directions：*You will be divided into three major groups, each of which will be assigned a topic. In each group, some students may be the manager, while others may be candidates. In the process of discussion, please observe the principles of cooperation, politeness and choice of words. One of the groups will be chosen to demonstrate the discussion to the class.*

1) Food advertising in our daily life.

2) The most impressive food advertising in your mind.

3) The key factors of a successful food advertisement.

Task Four   Discussion and Debate

Directions：*The class will be divided into two groups. Please choose your stand in regard to the following controversy and support your opinions with scientific evidences. Please refer to the specialized terms and classical sentences in the previous parts of this unit.*

Creation and innovation has always been treated by many as one of the overriding necessities and challenges for the development of any corporations. Some, however, advocate strongly that it is far from the truth in food industry. To them, keeping the original taste helps to attract and retain customers. What do you think? Please use examples to support your idea.

V. After-class Exercises

1. *Match the English words in Column A with the Chinese meaning in Column B.*

A	B
1) trademark	a. 库存

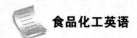

2）inventory   b. 商品组合

3）merchandising   c. 销售量

4）volume   d. 货架

5）assortment   e.（注册）商标

6）fabricated food   f. 购买惯性

7）gondola   g. 合成食品,组合食品

8）differentiated marketing   h. 助销

9）buying inertia   i. 顾客维系

10）customer retention   j. 差异化营销

2. *Fill in the following blanks with the words or phrases in the word bank. Some may be chosen more than once. Change the forms if it's necessary.*

fragmentation	sponsor	manufacturing	public speaking
newspapers	magazines	definitive	hyperactive
segmentation	nurture	elaborate	
coverage of the newspaper/its customer base/the cost of advertising			
unification	product	price	promotion
place	sound	sight	franchising

1）If the product you sell relates to your own expertise,＿＿＿ can be a great advertisement.

2）＿＿＿ and ＿＿＿ are the earliest forms of press advertising.

3）Before selecting a newspaper, you need to consider various factors such as ＿＿＿ etc.

4）Television commercial has the advantages of ＿＿＿ and ＿＿＿.

5）＿＿＿ is the pattern of the right to use an organization's business model and brand for a particular period of time.

6）There are three historical phases of food marketing: the ＿＿＿ phase (before 1870－1880), the ＿＿＿ phase (1880－1950), and the ＿＿＿ phase (1950 and later).

7）In the ＿＿＿ phase, the United States was divided into numerous geographic fragments.

8）In the unification phase, product consistency was made possible by advances in ＿＿＿.

9）In the ＿＿＿ phase (1950 and later), radio, television and internet advertising made the product competition more fierce.

10）The four components of food marketing relate to ＿＿＿, ＿＿＿, ＿＿＿ and ＿＿＿.

3. *Translate the following sentences into English.*

1）众所周知,公司并购中的税收问题是一个很复杂的课题。

＿＿＿＿＿＿＿＿＿＿＿＿＿＿＿＿＿＿＿＿＿＿＿＿＿＿＿＿＿＿＿

＿＿＿＿＿＿＿＿＿＿＿＿＿＿＿＿＿＿＿＿＿＿＿＿＿＿＿＿＿＿＿

2）含有添加剂的新产品都将会有相对较长的生命周期。

＿＿＿＿＿＿＿＿＿＿＿＿＿＿＿＿＿＿＿＿＿＿＿＿＿＿＿＿＿＿＿

3）常接触营销广告的儿童和青少年更容易肥胖。

_____

_____

4）在同一阶段，电报和电话提高了协调销售队伍的效率。

_____

_____

5）食品营销是让大众了解其产品/服务，对其感到兴奋并驱使他们购买。

_____

_____

4. *You are the sales manager of a milk company which has launched a new kind of milk into the market. The milk contains some useful nutrients for children. Now please write a direct email about 120 words to the parents in your community, your targeted customers, to introduce this product. Please try your best to make your customers excited and compel them to purchase it. Watch out the spelling of some specialized terms you have learnt in this unit.*

_____

_____

_____

_____

_____

_____

_____

_____

_____

_____

_____

_____

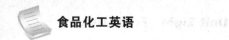

VI. Additional Reading

Burger King:Seeking Consistency in Leadership and Image

For nearly 60 years,Burger King has served flame-broiled (火焰焙烧的) hamburgers at an affordable price. In this sense,the fast-food chain best known for its over-sized sandwich has been nothing but consistent. Beyond the broiler,however,Burger King's most notable trait is its inconsistency. The corporation's longest CEO tenure was four years. When Brad Blum took over the position in 2004,he was the ninth to lead in 15 years. Since then,five others have held the job——the most recent being a former railroad executive with no fast-food experience.

Some say the overriding (最重要的) motivation of several owners was to deprive Burger King of capital in order to pay parent shareholders quickly. Others trace the chain's decline to the mid-90s,when then-owner Diageo PLC neglected the company and its tanking sales. Regardless of one's perspective,such a degree of ownership turnover leaves no time to develop a coherent strategy,let alone implement (执行) one.

It is advised to examine how inconsistency of leadership and its byproduct have overwhelmingly contributed to Burger King's repeated financial struggles. Now more than ever, as the chain launches its broadest menu expansion in its 58-years and makes its biggest communications push in its history,Burger King must establish effective leadership strategies and implement and diffuse (传播) a compelling vision. Addressing these issues will have a direct effect on its image,which will translate to consumers and increase profits.

Burger King's history

James McLamore and David Edgerton founded Burger King Corporation in 1954. The men had been Miami,Florida franchisees (特许经营人) of the restaurant's precursor,Insta-Burger King. Their business strategy was simple:attract postwar baby boomer families with reasonably priced, swiftly served broiled hamburgers. Their plan was not unique;however,McLamore and Edgerton set their chain apart by being the first to offer dining rooms and,in 1957,the Original Whopper Sandwich. A year after the iconic burger appeared on Burger King's first-ever television advertisements,McLamore and Edgerton expanded their five Florida restaurants into a nationwide enterprise. When Burger King was sold to the Pillsbury Company in 1967,it was the third largest fast-food chain in the United States with 275 locations.

Pillsbury owned Burger King for 30 years. Yet problems within the corporation's ranks quickly emerged. In 1970,Burger King battled with a pair of franchisees,Billy and Jimmy Trotter. The Trotters had bought up restaurants around the country before attempting to buy the chain for $100 million. Pillsbury refused and sued a group of Boston franchisees who had sold to the Trotters,claiming Pillsbury had the right to refuse the sale. The Trotters compromised and relinquished（放弃）their Boston holdings.

The end result reinforced Pillsbury's transactional（交易的）and authoritative（权威的）management style. It also paved a future path of contentious（有争论的）relationships with franchisees. Through its actions, Burger King advanced the interests of itself and individual storeowners, yet was not interested in a deep or enduring link. With the 1977 addition of Donald Smith, a hard-hitting executive from current number-one fast food chain McDonald's, Burger King introduced a more demanding franchisee contract which stipulated（规定）"that franchisees may not own other restaurants and must live within an hour's drive of their franchise". To many franchisees, the company became overbearing（专横的）, a negative aspect of the authoritative leadership style which can "undermine the egalitarian（主张平等的）spirit of an effective team". This feeling of franchisee oppression（压迫）has persisted throughout the company's history.

The need for change

Earlier this year, former CEO Brad Blum said："Burger King has got to focus on who they are and what sets them apart." Meeting this objective may be the goal of the company's new marketing campaign, which has been called its most ambitious. Before examining this campaign, it is important to review the main reasons why it and other organizational changes were necessary.

Leadership volatility

Burger King's fluctuating（波动的）leadership has forced the chain to engage in a dissonant（不和谐的）style of leadership, created when there is a lack of harmony between potential leaders and followers. While this lack of harmony is a natural consequence of frequent ownership changes, the company failed to understand its impact on franchisees and effectively manage change. Had Burger King taken a system theory approach, it may have understood that its long-term flagging sales were not independent of other parts of its system, namely how it treated its franchisees. Acknowledging this may have slowed the company from implementing shake-ups when new ownership took over. Instead, it moved "too quickly through tough steps such as developing a clear vision or an effective strategy". As a result, Burger King is an organization in constant flux（不断的变动）; that desperately needs stability.

Faltering image

In recent years, Burger King's marketing strategy was geared exclusively to those it termed "superfans"：18 to 35-year-old males. The company teamed with advertising agency and "creative powerhouse" Crispin Porter + Bogusky to create an edgy campaign strategy highlighted by television advertisements featuring a "creepy-looking king" and online gimmicks（花招）like the Subservient Chicken. For a short period of time, this "cool" strategy seemed to work. In February 2007, Burger King reported a 12th consecutive quarter of worldwide sales

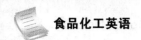

growth at stores open at least a year. However, Burger King failed to anticipate how the 2008 recession would affect its core demographic.

Failure to adapt to market

In 2010, industry analysts said Burger King's marketing and advertising focus alienated women, children and other customers. Yet executives within the company's ranks failed to notice. They also ignored the fact that from January to September 2009, 18- to 34-year olds went to fast-food chains about 13 times per month on average, down from almost 19 times a month in 2006. Even during the height of Burger King's coolness offensive, brand expert Martin Lindstrom questioned the chain's image, indicting its leadership: "It's still a slightly schizophrenic (精神分裂症的) brand."

Burger King's schizophrenia is evident in its marketing history. As far back as 1988, franchisees were dissatisfied with management's advertising strategies. Like its revolving door of CEOs, Burger King has had an array of marketing chiefs. In the late eighties, the company re-staffed the position twice in six months. It repeated history in 2011 with the dismissal (解雇) of chief marketing officer Natalia Franco after only nine months.

Burger King also failed to respond to the general demand for healthier eating options, an oversight its executives have readily admitted. In 2011, 9% of consumers said they wanted fresh and lighter fare, yet the company continued its focus on large portions for young males. According to Bonnie Riggs, a restaurant industry analyst at market research firm NPD Group, young people were actually looking for healthier options, or at least what they perceived to be healthier options. Overlooking this trend was a huge misstep, especially considering that of the 61 billion annual visits to restaurants in the United States, nearly one out of four is to a burger chain restaurant. While its competitors provided menu items in line with this movement, Burger King "failed to diversify its offerings" and "let its menu get stale".

What change is needed

Clearly, Burger King's culture of inconsistency is an epidemic that requires a cohesive and consistent change management effort by leadership. Analyst Denhardt, et al. wrote, "Three ideas seem central to bringing about change effectively. First, managers or other 'change agents' need to clarify and communicate the problems inherent in the current situation... Second, managers should involve people throughout the organization in the change process... Third, managers must recognize that people involved in change simply need time".

Use of various leadership styles

Change is often accompanied by conflict, so the former cannot be managed without addressing the latter. The two also go hand-in-hand with vision. Analyst Schein wrote, "In a rapidly changing world, the leader must not only have vision, but also be able to evolve it further as external circumstances change". As such, Burger King should heed (留心) the analyst' recommendation to practice collaborative behavior—the best way to manage and resolve conflict. All employees need to feel part of the change process, and the more they can contribute to the change discussions, the more they will start to internalize（使内在化）the proposed changes.

One way for Burger King to manage both change and conflict is through dialogic（对话体的）communication, which implies that relationships matter, different perspectives matter, and power is shared among all within the organization. This type of communication allows for engagement that demonstrates genuine care and respect and generates reflective discussion. Another way is to communicate with resonance（共鸣；反响）, not dissonance（不和谐）. Resonance enables the leader to use expertise in pursuit of a company's performance. It allows the leader to engage the power of all the people who work in and around the organization.

Emphasis on relationships within an organization and engaging employees are characteristics of the affiliative（有亲和力的）leadership style, which revolves around people and builds a sense of belonging. Sound relationships deepen a team's ability to think critically, resolve problems and innovate. This style also helps create team harmony and repair broken trust—goals Burger King has struggled to achieve.

Revamped menu and restaurant design

On April 2, 2012, Burger King rolled out 10 new items at once, an act never done before in the history of the company. Each item coincided（两件或更多的事情同时发生）with the chain's goal to offer healthier alternatives. Some of the new options were: homestyle chicken strips, chicken snack wraps, frappes（冰镇饮料）, smoothies, and salads. Many observers noted these items have been part of competitors' menus for years. Steve Wiborg, president of Burger King's North America operations, did not shy away from the accusation. "Consumers wanted more choices," he said, "not just healthy choices, but choices they could get at the competition."

The company also openly admitted that its new menu additions are aimed at a broader range of consumer groups. Burger King representatives did not simply walk into a McDonald's and write down all the items they wanted to sell. Company executives say the chain came up with them through extensive research, another indication that Burger King is taking the creation of its new image and vision seriously. Over the past year, Burger King evaluated all of its ingredients from the bacon to the cheese slices it serves on its chargrilled burgers as part of a yearlong quest to reverse years of slumping sales of its Whoppers and fries. Even seemingly straightforward items when through several variations before the final version was selected and months were spent determining the right mix for its new line of smoothies.

Executives also hosted focus groups, a popular method for receiving input from a large number of individuals. This method of consumer feedback has advantages over surveys in that participants tend to feel empowered（授权）and may query one another and explain their

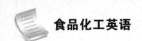

answers to others. Burger King officials found that the Whopper was still a customer favorite, but that the chain had to catch up in some important product categories, specifically salads, smoothies and wraps.

Jordan Krolick, president of the consulting firm Tound & Drowth who held senior positions at Arby's and McDonald's said: "You can change the menu and the advertising, but you're not going to get customers to see those changes without fresh, new, clean-looking facilities." To this end, Burger King is also in the process of modernizing 7 200 of its aging stores, redesigning worker uniforms to stay cleaner, and even serving the Whopper in cardboard cartons instead of paper wrapping for the first time in 20 years. The company is also digitizing its menu boards, making its chairs and tables mostly moveable instead of cemented to the ground, and adding chairs that feel more like couches in some stores. All in an effort to modernize its image.

Advertising

In March 2011, Burger King announced it was cutting ties with "cool" Crispin Porter + Bogusky in favor of the more product-focused services of New York-based advertising firm McGarryBowen. The swap (交换) signaled the end of "The King", the leering plastic-y mascot featured in most of its previous advertising efforts. McGarryBowen's plan was to concentrate on promotion of Burger King's actual food. In a serious image twist, many of Burger King's entire commercials show only the sights and sounds of the fresh ingredients being washed, sliced (切成片) and diced (切成丁).

Burger King also expanded its relationship with California ad agency Pitch by selecting it for promotions aimed at children and families over the company once in charge of these groups, Campbell Mithun. Both moves clearly demonstrate Burger King is abandoning its courtship of "young, hungry guys" for a campaign more culturally directed towards moms, families and boomers.

Burger King also enlisted some A-list celebrities to help spread the word about its new menu, including David Beckham, Jay Leno, Steven Tyler, Mary J. Blige and Selma Hayek. In the end, the ultimate measure of success will be whether or not these new methods positively impact the bottom line. Financial gains will not result from a poor image, poor franchisee relationships, and a poor product. These three factors are explicitly (明确地) and irrevocably (不能取消地) tied to profit, so Burger King must constantly—and consistently—monitor feedback and respond to concerns if the company wishes to reclaim the number two spot among North America's fast food chains. Then it must keep its eye on the broiler to remain there.

(*If you want to find more information about this corporation, please log on http://michaelpetitti. com/wp-content/uploads/2012/05/BurgerKingPETITTI.pdf*)

1. *Read the passage quickly by using the skills of skimming and scanning. And choose the best answer to the following questions.*

1) Burger King is inconsistent beyond _____.

    A. leadership   B. image

    C. over-sized sandwich   D. CEO tenure

2) According to the text, _____ is not one of Burger King's business strategy.

    A. reasonable price   B. community location

    C. swift service   D. being the first to offer dining rooms

3) The feeling of franchisee oppression is reflected in the fact that _____.

    A. franchisees may not own other restaurants

    B. franchisees must live within an hour's drive of their franchise

    C. Burger King refused and sued a group of Boston franchisees

    D. all of the above

4) In recent years, Burger King's marketing strategy had been geared exclusively to "superfans" of _____.

    A. 18- to 35-year-old males   B. 10- to 25-year-old females

    C. 15- to 30-year-old females   D. 18- to 28-year-old males

5) In 2010, industry analysts said Burger King's marketing and advertising focus alienated the following group except _____.

    A. women   B. other customers

    C. children   D. the elderly

6) Burger King also overlooked the trend of _____, while its competitors provided menu items in line with this movement.

    A. the general demand for healthier eating options

    B. consumers' need for swift service

    C. customer's sensitivity to advertising

    D. the effect of internet influentials

7) Facing changes, the best way to manage and resolve conflict for Burger King is to _____.

    A. let its menu get stale   B. diversify its offerings

    C. practice collaborative behavior   D. carry out monologue communication

8) On April 2, 2012, Burger King rolled out 10 new items at once except _____.

    A. decaf coffee   B. homestyle chicken strips

    C. chicken snack wraps   D. smoothies and salads

9) With an effort to modernize its image, Burger King _____.

    A. redesigned worker uniforms   B. served the Whopper in cardboard cartons

    C. digitized its menu boards   D. all of the above

10) From March 2011, Burger King did not have a cooperation with _____, the ad agency.

    A. Pitch   B. McGarryBowen

C. Crispin Porter +Bogusky D. Campbell Mithun

2. *In this part, you are required to make an oral presentation on either of the following topics.*

1) The SWOT analysis of Burger King.

2) The measures Burger King has taken to face changes.

习题答案

Unit Nine Application of Chemistry

I. Pre-class Activity

Directions：*Please read the general introduction about Alfred Bernhard Nobel and tell something more about the great scientist to your classmates.*

Alfred Bernhard Nobel

Alfred Bernhard Nobel was a Swedish chemist, engineer, inventor, businessman, and philanthropist.

Known for inventing dynamite, Nobel also owned Bofors, which he had redirected from its previous role as primarily an iron and steel producer to a major manufacturer of cannon and other armaments. Nobel held 355 different patents, dynamite being the most famous. After reading a premature obituary which condemned him for profiting from the sales of arms, he bequeathed his fortune to institute the Nobel Prizes. The synthetic element nobelium was named after him. His name also survives in modern-day companies such as Dynamit Nobel and Akzo Nobel, which are descendants of mergers with companies Nobel himself established.

Born in Stockholm, Alfred Nobel was the third son of Immanuel Nobel (1801–1872), an inventor and engineer, and Carolina Andriette (Ahlsell) Nobel (1805–1889). The couple married in 1827 and had eight children. The family was impoverished, and only Alfred and his three brothers survived past childhood. Through his father, Alfred Nobel was a descendant of the Swedish scientist Olaus Rudbeck (1630–1702), and in his turn the boy was interested in engineering, particularly explosives, learning the basic principles from his father at a young age. Alfred Nobel's interest in technology was inherited from his father, an alumnus of Royal Institute of Technology in Stockholm.

Following various business failures, Nobel's father moved to Saint Petersburg in 1837 and grew successful there as a manufacturer of machine tools and explosives. He invented modern

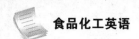

plywood and started work on the torpedo. In 1842, the family joined him in the city. Now prosperous, his parents were able to send Nobel to private tutors and the boy excelled in his studies, particularly in chemistry and languages, achieving fluency in English, French, German and Russian. For 18 months, from 1841 to 1842, Nobel went to the only school he ever attended as a child, the Jacobs Apologistic School in Stockholm.

As a young man, Nobel studied with chemist Nikolai Zinin; then, in 1850, went to Paris to further the work. There he met Ascanio Sobrero, who had invented nitroglycerin three years before. Sobrero strongly opposed the use of nitroglycerin, as it was unpredictable, exploding when subjected to heat or pressure. But Nobel became interested in finding a way to control and use nitroglycerin as a commercially usable explosive, as it had much more power than gunpowder. At age 18, he went to the United States for one year to study chemistry, working for a short period under inventor John Ericsson, who designed the American Civil War ironclad USS Monitor. Nobel filed his first patent, an English patent for a gas meter, in 1857, while his first Swedish patent, which he received in 1863, was on ways to prepare gunpowder.

The family factory produced armaments for the Crimean War (1853–1856), but had difficulty switching back to regular domestic production when the fighting ended and they filed for bankruptcy. In 1859, Nobel's father left his factory in the care of the second son, Ludvig Nobel (1831–1888), who greatly improved the business. Nobel and his parents returned to Sweden from Russia and Nobel devoted himself to the study of explosives, and especially to the safe manufacture and use of nitroglycerin. Nobel invented a detonator in 1863, and in 1865 designed the blasting cap.

On 3 September 1864, a shed used for preparation of nitroglycerin exploded at the factory in Heleneborg, Stockholm, killing five people, including Nobel's younger brother Emil. Dogged and unfazed by more minor accidents, Nobel went on to build further factories, focusing on improving the stability of the explosives he was developing. Nobel invented dynamite in 1867, a substance easier and safer to handle than the more unstable nitroglycerin. Dynamite was patented in the US and the UK and was used extensively in mining and the building of transport networks internationally. In 1875, Nobel invented gelignite, more stable and powerful than dynamite, and in 1887 patented ballistite, a predecessor of cordite.

Nobel was elected a member of the Royal Swedish Academy of Sciences in 1884, the same institution that would later select laureates for two of the Nobel prizes, and he received an honorary doctorate from Uppsala University in 1893.

II. Specialized Terms

Directions: *Please memorize the following specialized terms before the class so that you will be able to better cope with the coming tasks.*

aerogel n.气凝胶

aesthetic adj.美学的

agro-science n.农业科学

align v.排列

alkali ash 碱灰

antimicrobial adj.抗菌的

artery n.动脉

biodegradation n.生物降解

biodiversity n.生物多样性

biomedical n.生物医药

bitumen n.沥青

bleach n.漂白剂

body wash 沐浴露

brew v.酿造

bronchitis n.支气管炎

bullet-proof vest 防弹衣

cast-iron adj.铸铁的

catalytic converter 催化转化器

chip n.芯片

commodity n.商品

cosmetics n.化妆品

crust n.外壳

derivatization n.衍生作用

detox n.解毒

dice v.切成方块

diesel n.柴油车;柴油

digestible adj.易消化的

expire v.到期

fatty acid 脂肪酸

ferric adj.铁的

fertilizer n.肥料

glycerol n.甘油

grilled adj.烤的

habitat n.栖息地

hardness n.硬度

hazard n.危害

hydroscopic adj.吸湿的

instrumentation n.仪表

microscope n.显微镜

monoculture n.单种栽培

multimillion n.数百万

nanocoating n.纳米涂料

nanotechnology n.纳米技术

navigation n.导航

NGO 非政府组织

NFPA 美国消防协会

nitrate n.硝酸盐

nonpolar adj.无极的

nucleotide n.核苷酸

oxidizer n.氧化剂

pesticide n.杀虫剂,农药

petroleum n.石油

pollen n.花粉

pollutant n.污染物

pool n.物资

porous adj.多孔的

portfolio n.产品组合

pose v.造成

potent adj.强有力的

primarily adv.主要地

proactively adv.主动地

quadrillion n.千万亿

quadruple v.使……成四倍

quench v.解渴

quotation n.报价

receipt n.收据

refrigerant n.冷冻剂

reproductive adj.生殖的

restore v.恢复

ripen v.成熟

roast v.烤

rotation n.旋转

semiconductor n.半导体

shovel n.铲子

slice v.把……切成片

specification n.规格；详述；说明书

squarely adv.直接地

swirl v.旋转

switch v.转变

temperament n.气质

textile n.纺织品

thermostats n.恒温器

tin n.罐

triglyceride n.甘油三酯

ultimate adj.最终的

viable adj.可行的

virtually adv.实质上

wafer n.晶片

waste disposal equipment 废物处理设备

watch for 提防

waterborne adj.水传播的；水运的

weaning food 断奶食品，离乳食品

weld v 焊接

well-earned adj.应得的

winnow v.风选(以去掉谷壳)；筛选

winnower n.风筛器，扬壳机

wither v.衰弱

workup n.检查

ziplock bag 自封袋

III. Watching and Listening

Task One　Six Chemical Reactions that Changed History

New words

brown v.变成褐色

hysteria n.歇斯底里

drinkable water 可饮用水

tablet n.碑

smelly adj.发臭的

toga n.托加长袍(古罗马服饰)

视频链接及文本

Exercises

1. *Watch the video for the first time and choose the best answers to the following questions.*

　　1) Which benefit of fire is not mentioned?

　　　　A. More nutrition.　　　　　　　B. Less work.

　　　　C. Easier to digest.　　　　　　D. Less bacteria.

　　2) _____ was the beginning of human's heavy metal stage.

　　　　A. Iron　　　　　　　　　　　　B. Bronze

　　　　C. Gold　　　　　　　　　　　　D. Platinum

　　3) What is not the form that sugar is converted into?

　　　　A. Acid.　　　　　　　　　　　　B. Alcohol.

　　　　C. Gas.　　　　　　　　　　　　D. Solid.

　　4) 4000 years ago, what is not the ingredient for making soap?

　　　　A. Animal fat.　　　　　　　　　B. Water.

　　　　C. Acid.　　　　　　　　　　　　D. Oil.

　　5) For most of life history, converting nitrogen to biological useful form could only be done by bacteria in _____.

A. soil   B. air

C. river   D. Ocean

2. *Watch the video again and decide whether the following statements are true or false.*

1) It is believed that Maillard reaction is the most delicious reaction in human history.(   )

2) Bronze is a step up in hardness and durability than pure copper.(   )

3) As the poet John Ciardi put it："Fermentation and civilization are separable."(   )

4) Only animal fat are triglycerides.(   )

5) Neither cell phones nor smart thermostats would be possible without silicon chips.(   )

3. *Watch the video for the third time and fill in the following blanks.*

For the first time, farmers didn't have to rely on crop 1) _____ or shovel what the family cow provided them to get 2) _____. Inexpensive chemical fertilizers let many people grow 3) _____ food for the first time ever. The world grew so much food, in fact, that the global population has 4) _____ since this chemical reaction. We make 5) _____ million tons of nitrogen fertilizer this way every year, a full one to two percent of all the 6) _____ we use goes to that process. Of course, salad bars or cereal aren't the only thing that we make with 7) _____ nitrogen. Nitrates are necessary ingredients for making 8) _____.

4. *Share your opinions with your partners on the following topics for discussion.*

1) What do you think is the most important chemical reaction? Why do you think so?

2) What kind of chemical reaction do you look forward to carrying out?

Task Two   Chemistry-inspired Tricks

New Words

skillet n.长柄平底煎锅   tarnish n.污点

sink n.水槽   blackish adj.带黑色的

Brillo Pad 美国一种细毛刷子的商标名

视频链接及文本

Exercises

1. *Watch the video for the first time and choose the best answers to the following questions.*

1) What makes the coffee less bitter?

A. Ginger.   B. Salt.

C. Garlic.   D. Onion.

2) What makes fruit ripen faster?

A. Ethylene.   B. Propylene.

C. Methane.   D. Ethane.

3) What is known to be hygroscopic?

A. Salt.   B. Pepper.

C. Vinegar.   D. Sugar.

4）What is the original function of phosphoric acid?

    A. Remove rust and tarnish.   B. Make it sweet.

    C. Make it bitter.   D. Add a sour punch.

5）Phosphoric acid can convert your typical had-to-clean iron oxide rust into _____.

    A. phosphate   B. potassium phosphate

    C. ferric phosphate   D. sodium phosphate

2. *Watch the video again and decide whether the following statements are true or false.*

  1）The hotter the water, the deeper the extraction from your beans and the more bitter compounds end up in your brew.（   ）

  2）Sodium ions block bitter molecules from reaching your tongue.（   ）

  3）The riper the banana get, the more ethane they produce.（   ）

  4）The bananas, depending on how green they were to start, will ripe twice as fast.（   ）

  5）Bread contains far more sugar than cookies.（   ）

3. *Watch the video for the third time and fill in the following blanks of the table.*

Chemical reactions	Problem	Solution	Mechanism
1）Coffee cream & salt	Coffee is too bitter		When salt dissolve, sodium ion break off into your coffee and block bitter molecules from reaching your tongue
2）Brown bagging bananas		Put bananas and ripe tomatoes in a bag	
3）How to save a cookie			
4）Iron and coke			

4. *Share your opinions with your partners on the following topics for discussion.*

  1）Do you have any tricks to overcome difficulties? Please share ideas with your classmates.

  2）Are there any taboos in mixing food? Try to explain one from the perspective of chemical reactions.

IV. Talking

Task One　Classical Sentences

Directions：*In this section, some popular sentences are supplied for you to read and to memorize. Then, you are required to simulate and produce your own sentences with reference to the structure.*

General Sentences

  1. My hobby is collecting stamps. Do you have a hobby?

我的爱好是集邮。你有什么爱好？

2. I've always thought photography would be an interesting hobby.
我一直认为摄影是一种有趣的爱好。

3. Some people like horseback riding, but I prefer golfing as a hobby.
一些人喜欢骑马,但是我喜欢打高尔夫。

4. Do you have any special interests other than your job?
除了工作以外,你还有什么其他特殊爱好吗？

5. Learning foreign languages is just an avocation with me.
学外语只是我的业余爱好。

6. I find stamp collecting relaxing and it takes my mind off my work.
我发现集邮使人放松,能让我的注意力从工作中转移。

7. On weekends I like to get my mind off my work by reading good books.
周末我喜欢通过读好书来把我的注意力从工作上转移。

8. My cousin is a member of a drama club. He seems to enjoy acting.
我堂兄是戏剧俱乐部的成员,他似乎喜欢表演。

9. He plays the piano for his own enjoyment.
他弹钢琴是为了自娱自乐。

10. I've gotten interested in Wi-Fi. I'm building my own equipment.
我对无线网络很感兴趣,我正在安装我自己的设备。

11. He's not a professional. He plays the piano for the fun of it.
他不是专业人士,他弹钢琴是为了好玩。

12. I've heard of unusual hobbies, but I've never heard of that one.
我听说过一些不寻常的爱好,但我从来没听说过那一个。

13. The trouble with photography is that it's an expensive hobby.
摄影的问题在于,它是一种昂贵的爱好。

14. That's a rare set of coins. How long did it take you to collect them?
这是一套罕见的钱币。你用了多久收集的？

15. I started a new hobby. I got tired of working in the garden.
我开始了一个新的爱好,我厌倦了在花园里干活。

16. Baseball is my favorite sport. What's your favorite?
棒球是我最喜欢的体育运动。你最喜欢的是什么？

17. My nephew is a baseball player. He is a catcher.
我的外甥是一名棒球运动员,他是一名接球手。

18. When you played football, what position did you play?
你踢足球时,踢什么位置？

19. We played a game last night. The score was six-to-six.
我们昨晚玩了一场比赛,比分是 6 比 6 平。

20. I went to a boxing match last night. It was a good fight.

昨晚我去看了一场拳击比赛,比赛很精彩。

21. When I was on the track team, I used to run the quarter mile.
我在田径队时,经常跑四分之一英里。

22. I like fishing and hunting, but I don't like swimming.
我喜欢钓鱼和打猎,但是不喜欢游泳。

23. My favorite winter sport is skiing. I belong to a ski club.
我最喜欢的冬季运动是滑雪。我是滑雪俱乐部的成员。

24. Would you be interested in going to the horse races this afternoon?
今天下午你有兴趣去看赛马吗?

25. What's your favorite kind of music? Do you like jazz?
你最喜欢什么音乐? 你喜欢爵士乐吗?

26. He's a composer of serious music. I like his music a lot.
他是一名严肃音乐作曲家,我很喜欢他的音乐。

27. My brother took lessons on the trumpet for nearly ten years.
我的哥哥练习吹喇叭将近十年了。

28. You play the piano beautifully. How many hours do you practice every day?
你钢琴弹得很好,你一天练习多长时间?

29. I've never heard that piece before. Who wrote it?
我从没有听过这一段,是谁写的?

30. Have you ever thought about becoming a professional musician?
你有没有想到过要成为一名职业音乐人?

31. Be a good sport. Play according to the rules of the game.
做个有风度的人,遵守游戏规则。

32. Our family went camping last summer. We had to buy a new tent.
我家去年夏天去露营了。我们不得不买了新的帐篷。

33. This afternoon we went to the gym for a workout. We lifted weights.
今天下午我们去体育馆健身了,我们练了举重。

34. What do you do for recreation? Do you have a hobby?
你闲暇时做什么? 你有什么爱好吗?

35. My muscles are sore from playing baseball.
打完棒球后,我的肌肉一直酸痛。

36. I sent in a subscription to that magazine. It's put out every week.
我订阅了那份杂志,它是周刊。

37. If you subscribe to the newspaper, it'll be delivered to your door.
如果你订阅报纸的话,可以送到你家。

38. I didn't read the whole paper. I just glanced at the headlines.
我没有通读全文,我只是看了看标题。

39. The first chapter of the story is in this issue of the magazine.

这个故事的第一章刊登在本期杂志上。

40. I haven't seen the latest issue of the magazine. Is it out yet?

我还没有看到这个杂志的最新一期,是不是还没有出版?

41. What's the total circulation of this newspaper?

这个报纸的总发行量怎样?

42. I'm looking for the classified section. Have you seen it?

我在找分类广告栏。你看到了吗?

43. My brother-in-law is a reporter on *The New York Times* staff.

我姐夫是《纽约时报》的记者。

44. There was an article in today's paper about the election.

今天的报纸上有选举的消息。

45. There wasn't much news in the paper today.

今天的报纸上没有太多的消息。

46. How long have you been taking this magazine?

你订这份杂志多久了?

47. Did you read the article about the rescue of the two fishermen?

你读了那篇关于营救两名渔夫的文章了吗?

48. Why don't you put an advertisement in the paper to sell your car?

你为什么不在报纸上登个广告卖你的车呢?

49. I got four replies to my ad about the bicycle for sale.

我的自行车待售广告有了四个回复。

50. My son has a newspaper route. He delivers the morning paper.

我儿子有送报的路线,他送晨报。

Specialized Sentences

1. All samples of matter have two things in common.

所有物质的样本都有两个共同点。

2. You are involved with chemicals and chemical reactions every instant of your life.

生活中的任何瞬间,你都无法摆脱化学物和化学反应。

3. It requires a deep-going analysis.

这需要深入分析。

4. Home security systems protect people from damage by fire, explosion and toxic gases.

家庭安全系统保护人们不受火灾、爆炸和有毒气体的伤害。

5. Efforts have been made for the reduction of the erroneous detection.

人们努力减少错误的检测。

6. For a comfortable life at home, both temperature and humidity in the rooms should be controlled at some levels.

为了创造家里舒适的生活,房间里的温度和湿度应该控制在适当水平。

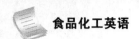
7. Many citizens would like to maintain full activity with good health throughout their lives.

很多人都希望一辈子充满活力、永远健康。

8. Chemical engineering has long been involved in the technologies used to convert natural resources into energy and useful products.

将自然资源转换成能源和有用产品的技术长期以来都离不开化学工程。

9. Increase in temperature favors this reaction.

温度的升高会促进该反应。

10. Chemical engineers apply scientific knowledge and technology for the betterment of humanity.

化学工程师运用科学知识和技术以改善民生。

11. Many technologies have found uses in more applications.

很多技术都有用武之地。

12. The 20th century saw dramatic advances in chemical application.

20 世纪见证了化学应用的巨大进步。

13. New concepts made chemical products an economic reality.

新概念把化学产品变成了经济现实。

14. Today chemical products become so common that we hardly notice them.

现如今,化学产品如此常见以至于我们几乎没有察觉。

15. All aspects of modern life are positively impacted by chemistry.

现代生活的方方面面都受到化学的积极影响。

16. Professors work hand in hand to help us live longer and better lives.

教授们联手合作,帮助我们提高寿命和生活质量。

17. Industrial creativity is a characteristic of chemical engineering.

化学工程的一大特点就是工业创新。

18. Raincoat keeps people out of harm's way.

雨衣保护人们免受伤害。

19. Chemistry provides effective answers to solve today's problems.

化学为解决当前的问题提供有效的方案。

20. Chemical application helps reduce the strain on nature.

化学应用帮助减少对自然界的压力。

21. Increase in technology offer the potential to increase the production.

技术进步使得提高生产成为可能。

22. Chemistry plays a prominent role in improving today's life.

化学在提高当今生活质量方面扮演重要的角色。

23. Some ingredients of make-up are toxic chemicals that may be hazardous to your body.

化妆品里的一些成分是有毒化学品,可能对身体有害。

24. Continued use of a product may not take effects after several months.

连续几个月持续使用一种产品,效果就不会那么明显。

25. Hazardous chemicals may cause body irritation.

有毒化学品会刺激人体。

26. Vitamin A encourages the generation of collagen, provides essential energy and improves the flexibility of skin.

维生素 A 促进胶原的产生,提供基本的能量以及提高皮肤的弹性。

27. It is especially recommended for use by children.

它特别适合儿童使用。

28. Since the industrial revolution, mankind has been dependent on fuel.

自从工业革命以来,人类就非常依赖燃料。

29. There will be an increasing reliance on clean energy.

人们将越来越依赖清洁能源。

30. The high temperature and pressure could be of particular importance in chemical reactions.

高温和高压对于化学反应来说,特别重要。

31. Progress in the chemical industry places greater emphasis on product design.

化学产业的进步会更加强调产品的设计。

32. Some special products are an important focus of the present chemical industry.

一些特殊产品是当今化学产业的重点。

33. We divide chemical products into three categories.

我们将化学产品分为三类。

34. These four products have nothing in common.

这四个产品毫无共同之处。

35. We begin by specifying what is needed.

我们开始就要清楚需要什么。

36. The entire effort is best viewed as a whole.

所有的努力都密不可分。

37. The key of process design is efficient manufacture.

产品设计过程的关键是高效的生产。

38. Solar power can be an environmentally friendly alternative to fuel nowadays.

当今,太阳能可以是燃料的环保替代品。

39. Its little weight has raised high expectations in modern science.

因为它重量轻,人们在当代科学中赋予它很高的期待。

40. We have developed strong partnerships with some of the most prominent names in this field.

我们已经和这个领域中的一些知名的品牌发展稳固的合作伙伴关系。

41. Expertise and innovation create solutions that add real value for you.

专业知识和创新为你提供了提升价值的解决方案。

42. Investing in the latest technology means we are often the first to market products and

services.

对最新技术的投资意味着,我们往往是第一批营销产品和服务的先锋者。

43. Manufacturers need to make sure our products match the needs of customers.

生产商需要确保我们的产品满足顾客的需要。

44. Machines rely on certain chemicals to get the best performance.

机器依赖某些化学品,以获得最佳性能。

45. It is critical that these machines stay in top condition.

这些机器保持最佳状态,非常重要。

46. Ensuring the safety of products is a responsibility that every brand takes very seriously.

保证产品的安全是每一个品牌需要认真考虑的责任。

47. We aim at providing branded products and services of superior quality and value that improve the lives of the world's consumers.

我们旨在提供具有高质量和价值的品牌产品和服务,以提高全世界消费者的生活水平。

48. What we need to do is to develop the capability to deliver our strategies and achieve objectives.

我们需要做的就是发展落实策略和实现目标的能力。

49. It combines the power of science and technology to innovate what is essential to human progress.

它将科学和技术的力量融合起来,以对人类进步的必经之路进行创新。

50. This company operates its worldwide operations through global business.

这个公司在全球范围内开展业务。

Task Two   Sample Dialogue

Directions: *In this section, you are going to read several times the following sample dialogue about the relevant topic. Please pay special attention to five C's (culture, context, coherence, cohesion and critique) in the dialogue and get ready for a smooth communication in the coming task.*

Introducing a skin care product

Sally: Good morning, what's wrong with your face? It looks so red.

Sophia: Oh, I am allergic to the skin care product that I just bought last week.

Sally: I'm sorry to hear that. Is it due to the quality of products or your sensitive skin?

Sophia: It is my skin that is very sensitive to cosmetic products. So I do not like to put chemicals on my face.

Sally: Well, it is hard for girls not to use any cosmetic products, especially in summer. There is no doubt that we need to protect ourselves from sunshine, dust and radiation.

Sophia: The product that I applied to my face is a world-famous brand. I am wondering why I am still allergic to it.

Sally: Actually, many world-famous brands are designed and sold for Americans and Europeans, which do not fit Asians.

Sophia: In this way, what do you advocate?

Sally: I prefer domestically made products like Pehchaolin.

Sophia: I always believe that price is positively related to the quality of products. If it is expensive, then the product is of high quality. Vice versa.

Sally: This rule doesn't work in every field.

Sophia: Domestically made products are designed exclusively for us Chinese, therefore it is very suitable for us to use them.

Sally: It makes sense. It is really a good deal if I can get a good quality product at such a low price. Thank you for your advice.

Sophia: My pleasure.

Task Three   Simulation and Reproduction

Directions: *The class will be divided into three major groups, each of which will be assigned a topic. In the process of discussion, please observe the principles of cooperation, politeness and choice of words. One of the groups will be chosen to demonstrate the discussion to the class.*

1) Hand cream

2) Serum

3) Moisturizer

Task Four   Discussion and Debate

Directions: *The class will be divided into two groups. Please choose your stand in regard to the following controversy and support your opinions with scientific evidences. Please refer to the specialized terms and classical sentences in the previous parts of this unit.*

Nowadays, foreign brands gain great popularity all over the world. It is believed that those branded products can provide supreme quality and service. However, an increasing number of users find that domestically made products are more suitable at lower cost. They think that personal experience ranks first among other factors. Which party do you agree with? Why?

V. After-class Exercises

1. *Match the English words in Column A with the Chinese meaning in Column B.*

A	B
1) pesticide	a. 耐用性
2) biodegradation	b. 危害

3) chip   c. 乙烯

4) hazard   d. 芯片

5) reactor   e. 硝酸盐

6) disinfectant   f. 生物降解

7) nitrate   g. 消毒剂

8) durability   h. 杀虫剂

9) ferric   i. 吸湿的

10) hygroscopic   j. 反应器

11) ethylene   k. 铁的

2. *Fill in the following blanks with the words in the word bank. Change the forms if it's necessary.*

adhesive	monoculture	antibiotic	catalyst	pesticide
detergent	disinfectant	fertilizer	hormone	phosphate
tarnish	reactor	yield	textile	viable

1) Basic chemicals include polymers, petrochemicals, reaction intermediates, inorganic chemicals and _____.

2) Fertilizer belongs to the smaller category of basic chemicals and includes _____, ammonia and potash.

3) Specialty chemicals include special _____, electronic chemicals, industrial gases and coatings.

4) Consumer chemicals include such things as soap, _____ and cosmetic products.

5) A number of review articles have identified subcritical water as an effective solvent, _____ and reactant for hydrolytic conversations and extractions.

6) Such operations might consist of heat exchangers, filters, chemical _____ and the like.

7) Chemical engineers have been able to take small amounts of _____ developed by people like Sir Authur Fleming.

8) They increase their _____ several thousand times through mutation.

9) DEA is a _____ disruptor and robs the body of choline needed for fetal brain development.

10) Formaldehyde is used as a _____ and preservative in a variety of products.

3. *Translate the following paragraph into Chinese.*

  Research into energy sources remains a key issue. Over the last 80 years, fuel has been the leading source of primary research work in fuel science. The scope is broad and includes many topics of increasing interest, such as environmental pollution.

_____

_____

_____

4. *Translate the following paragraph into English.*

酸雨是当前常见的一个化学现象。这个现象在一些环境恶劣的地区经常出现,因此我们对此并不陌生。什么是酸雨呢?酸雨就是酸度较高的雨,其 pH 值低于 5。这样的雨水在日常的生活中会给人们的生活带来非常大的困扰,因为其不仅损害环境,还会对人体造成伤害。

5. *Please write a specification for a personal care product. Pay attention to the form and watch out the spelling of some specialized terms you have learnt in this unit.*

VI. Additional Reading

Brief Introduction on Bayer

Bayer is a German multinational pharmaceutical(制药的) and life sciences company. It is

headquartered in Leverkusen, where its illuminated(被照明的) sign is a landmark. Bayer's primary areas of business include human and veterinary(兽医的) pharmaceuticals, consumer healthcare products, agricultural chemicals and biotechnology products, and high value polymers(聚合物). The company is a component of the Euro Stoxx 50 stock market index. The company's motto is "science for a better life".

Bayer's first and best known product was aspirin; there is a dispute about which scientist at Bayer made the most important contributions to it, Arthur Eichengrün or Felix Hoffmann. Bayer trademarked the name "heroin" for the drug diacetylmorphine(二乙酰吗啡) and marketed it as a cough suppressant and non-addictive substitute(替代物) for morphine from 1898 to 1910. Bayer also introduced phenobarbital(苯巴比妥,一种镇静安眠剂), prontosil(百浪多息), the first widely used antibiotic and the subject of the 1939 Nobel Prize in Medicine, the antibiotic Cipro (ciprofloxacin,环丙沙星), and Yaz (drospirenone,屈螺酮) birth control pills. In 2014, Bayer bought MSD's consumer business, with brands such as Claritin, Coppertone and Dr. Scholl's. Its Bayer Crop Science business develops genetically modified crops and pesticides.

Bayer was founded in Barmen in 1863. It was part of IG Farben, the world's largest chemical and pharmaceutical company, from 1925 to 1952, and became an independent company again after IG Farben was broken up by the Allies after World War Ⅱ due to its collaboration with the Nazi regime(政权). The company played a key role in the Wirtschaftswunder during the early Cold War and quickly regained its position as one of the world's largest chemical and pharmaceutical corporations. Bayer acquired Schering in 2006 and announced its acquisition of Monsanto in 2016.

Bayer AG was founded in Barmen (today a part of Wuppertal), Germany in 1863 by Friedrich Bayer and his partner, Johann Friedrich Weskott. Corresponding to his education Friedrich Bayer was responsible for commercial tasks at the young company. Weskott had an apprenticeship(学徒身份) as dyer(染工), which gave him knowledge in chemistry. Fuchsine and aniline became the most important products of the company.

The company's corporate logo, the Bayer cross, was introduced in 1904. It consists of the horizontal word "BAYER" crossed with the vertical word "BAYER", both words sharing the "Y", and enclosed in a circle. An illuminated version of the logo is a landmark in Leverkusen, the location of Bayer AG's headquarters.

During World War I, Bayer assets, including the rights to its name and trademarks, were confiscated(没收) in the United States, Canada, and several other

countries. In the United States and Canada, Bayer's assets and trademarks, including the well-known Bayer cross were acquired by Sterling Drug, a predecessor of Sterling Winthrop and wouldn't be reclaimed until 1994.

In 1916, Bayer scientists discovered suramin(苏拉明), an anti-parasite(抗寄生虫) drug that is still sold by Bayer under the brand name Germanin. The formula of suramin was kept secret by Bayer for commercial reasons; however, it was elucidated(阐明) and published in 1924 by Ernest Fourneau and his team of the Pasteur Institute. It is on the World Health Organization's List of Essential Medicines.

Bayer became part of IG Farben, a German chemical company conglomerate(企业集团), in 1925. In the 1930s, IG Farben scientists Gerhard Domagk, Fritz Mietzsch, and Joseph Klarer, discovered prontosil, the first commercially available antibacterial drug. The discovery and development of this first sulfonamide drug opened a new era in medicine. Domagk received the Nobel Prize in Medicine for this work in 1939.

During World War II, IG Farben used slave labor in factories that it built adjacent(邻近的) to German concentration camps, notably Auschwitz, and the sub-camps(分营) of the Mauthausen-Gusen concentration camp. IG Farben purchased prisoners for human experimentation of a sleep-inducing drug and later reported that all test subjects died. IG Farben

employees frequently said, "If you don't work faster, you'll be gassed." IG Farben held a large investment in Degesch which produced Zyklon B used to gas and kill prisoners during the Holocaust(大屠杀).

After World War II, the Allies broke up IG Farben and Bayer reappeared as an individual business "inheriting" many of IG Farben's assets. Fritz ter Meer, an IG Farben board member from 1926 to 1945 who directed operations at the IG Farben plant at Auschwitz, was sentenced to seven years in prison during the IG Farben Military Tribunal at Nuremberg. He was elected Bayer's supervisory board head in 1956.

In 1953, Bayer brought the first neuroleptic(安定精神的) Chlorpromazine onto the German market.

In the 1960s, Bayer introduced a pregnancy test, Primodos, that consisted of two pills that contained norethisterone(炔诺酮) and ethinylestradiol(乙炔雌二醇). It detected pregnancy by inducing menstruation(月经) in women who were not pregnant. The presence or absence of menstrual(月经的) bleeding was then used to determine whether the user was pregnant. The test became the subject of controversy when it was blamed for birth defects, and it was withdrawn from the market in the mid-1970s. Litigation(诉讼) in the 1980s regarding these claims ended inconclusively(不确定地). A review of the matter by the Medicines and

Healthcare Products Regulatory Agency in 2014 assessed the studies performed to date, and decided that it found the evidence for adverse effects to be inconclusive.

In 1978, Bayer purchased Miles Laboratories and its subsidiaries Miles Canada and Cutter Laboratories, acquiring along with them a variety of product lines including Alka-Seltzer, Flintstones vitamins and One-A-Day vitamins, and Cutter insect repellent(拒虫剂).

Along with the purchase of Cutter, Bayer acquired Cutter's Factor Ⅷ business. Factor Ⅷ, a clotting(凝固) agent used to treat hemophilia(血友病), was produced, at the time, by processing donated blood. In the early days of the AIDS epidemic(流行病), people with hemophilia were found to have higher rates of AIDS, and by 1983 the CDC had identified contaminated blood products as a source of infection. According to *the New York Times*, this was "one of the worst drug-related medical disasters in history". Companies, including Bayer, developed new ways to treat donated blood with heat to decontaminate it, and these new products were introduced early in 1984. In 1997, Bayer and the other three makers of such blood products agreed to pay $660 million to settle cases on behalf of more than 6 000 hemophiliacs infected in the United States. But in 2003, documents emerged showing that Cutter had continued to sell unheated blood products in markets outside the US until 1985.

In 1994, Bayer AG purchased Sterling Winthrop's over-the-counter drug business from SmithKline Beecham and merged it with Miles Laboratories, thereby reclaiming the US and Canadian trademark rights to "Bayer" and the Bayer cross, as well as the ownership of the Aspirin trademark in Canada.

In the late 1990s, Bayer introduced a statin(胆固醇合成酶抑制剂) drug, Baycol (Cerivastatin), but after 52 deaths were attributed to it, Bayer discontinued it in 2001. The side effect was rhabdomyolysis(横纹肌溶解), causing renal(肾脏的) failure, which occurred with a tenfold(十倍的) greater frequency in patients treated with Baycol in comparison to those prescribed alternate medications of the statin class.

Bayer has been involved in controversies regarding some of its other drug products. Trasylol (aprotinin), used to control bleeding during major surgery, was withdrawn from the markets worldwide when reports of increased mortality emerged; it was later reintroduced in Europe but not in the US.

In 2004, Bayer HealthCare AG (de) acquired the over-the-counter (OTC) Pharmaceutical Division of Roche Pharmaceuticals.

In March 2008, Bayer HealthCare announced an agreement to acquire the portfolio and OTC division of privately owned Sagmel, Inc., a US-based company that markets OTC medications in most of the Commonwealth of Independent States countries such as Russia, Ukraine, Kazakhstan, Belarus, and others. On 2 November 2010, Bayer AG signed an agreement to buy Auckland-based animal health company Bomac Group.

Bayer partnered on the development of the radiotherapeutic(放射疗法的) Xofigo with Algeta, and in 2014 moved to acquire the company for about $2.9 billion. In 2014, Bayer agreed to buy Merck's consumer health business for $14.2 billion which would provide Bayer control with brands such as Claritin, Coppertone and Dr. Scholl's. Bayer would attain second place globally in nonprescription drugs. In June 2015, Bayer agreed to sell its diabetic care business to Panasonic Healthcare Holdings for a fee of $1.02 billion.

In September 2015, Bayer spun out its $12.3 billion materials science division into a separate, publicly traded company called Covestro in which it retained about a 70% interest. Bayer spun out the division because it had relatively low profit margins compared to its life science divisions (10.2%, compared with 24.9% for the agriculture business and 27.5% for healthcare) and because the business required high levels of investment to maintain its growth, and to more clearly focus its efforts and identity in the life sciences. Covestro shares were first offered on the Frankfurt Stock Exchange in October 2015.

Effective January 2016 following the spinout (分拆公司) of Covestro, Bayer rebranded itself as a life sciences company, and restructured into three divisions and one business unit: Pharmaceuticals, Consumer Health, Crop Science, and Animal Health.

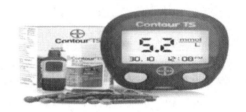

Bayer Crop Science has products in crop protection(i. e. pesticides), nonagricultural pest control, seeds and plant biotechnology. In addition to conventional agrochemical business, it is involved in genetic engineering of food. In 2002, Bayer AG acquired Aventis (now part of Sanofi) Crop Science and fused it with their own agrochemicals division (Bayer Pflanzenschutz or "Crop Protection") to form Bayer Crop Science; the Belgian biotech company Plant Genetic Systems became part of Bayer through the Aventis acquisition. Also in 2002, Bayer AG acquired the Dutch seed company Nunhems, which at the time was one of the world's top five seed companies. In 2006, the US Department of Agriculture announced that Bayer Crop Science's LibertyLink genetically modified rice had contaminated(把……弄脏) the US rice supply. Shortly after the public learned of the contamination, the EU banned imports of US long-grain rice and the futures price plunged(暴跌). In April 2010, a Lonoke County, Arkansas jury awarded a dozen farmers $48 million. The case is currently on appeal to the Arkansas Supreme Court. On 1 July 2011, Bayer Crop Science agreed to a global settlement for up to $750 million. In September 2014, the firm announced plans to invest $1 billion in the United States between 2013 and 2016. A Bayer spokesperson said that the largest investments will be made to expand the production of its herbicide(除草剂) Liberty. Liberty is used to kill weeds which have grown resistant to Monsanto's product Roundup. In 2016, as part of the wholesale corporate restructuring, Bayer Crop Science became one of the three major divisions of Bayer AG, reporting directly to the head of the division, Liam Condon.

Before the 2016 restructuring, Bayer HealthCare comprises a further four subdivisions(细

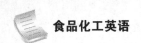

分）：Bayer Schering Pharma, Bayer Consumer Care, Bayer Animal Health and Bayer Medical Care. As part of the corporate restructuring, Animal Health was moved into its own business unit, leaving the division with the following categories: Allergy（过敏性反应）, Analgesics（镇痛剂）, Cardiovascular（心血管的）Risk Prevention, Cough & Cold, Dermatology（皮肤病学）, Foot Care, Gastrointestinals（胃肠的）, Nutritionals and Sun Care.

The Pharmaceuticals Division focuses on prescription products, especially for women's healthcare and cardiology（心脏病学）, and also on specialty therapeutics in the areas of oncology（肿瘤学）, hematology（血液学）and ophthalmology（眼科学）. The division also comprises the Radiology Business Unit which markets contrast-enhanced diagnostic imaging equipment together with the necessary contrast agents.

In addition to internal R&D, Bayer has participated in public-private partnerships. One example in the area of non-clinical safety assessment is the InnoMed PredTox program. Another is the Innovative Medicines Initiative of EFPIA and the European Commission.

Bayer HealthCare's Animal Health Division is the maker of Advantage Multi（imidacloprid + moxidectin）Topical Solution for dogs and cats, Advantage flea（跳蚤）control for cats and dogs and K9 Advantix, a flea, tick, and mosquito control product for dogs. Advantage Multi, K9 Advantix and Advantage are trademarks of Bayer. The division specializes in parasite control and prescription pharmaceuticals for dogs, cats, horses, and cattle. North American operation for the Animal Health Division are headquartered in Shawnee, Kansas. Bayer Animal Health is a division of Bayer HealthCare LLC.

In 2014, pharmaceutical products contributed €12.05 billion of Bayer's €40.15 billion in gross revenue. Top-selling products included:

Kogenate is a recombinant version of clotting factor Ⅷ, the absence of deficiency of which causes the abnormal bleeding associated with haemophilia type A.

Xarelto（rivaroxaban）is a small molecule inhibitor of Factor Xa, a key enzyme involved in blood coagulation.

Betaseron（重组干扰素 β-1b 生物制剂）is an injectable form of the protein interferon（干扰素）beta（贝塔,希腊字母中的第二个字母）used to prevent relapses（复发）in the relapsing remitting form of multiple sclerosis（多发性硬化）.

Yasmin（雅司明）birth control pills are part of a group of birth control pill products based on the progestin（孕激素）drospirenone.

Nexavar（sorafenib,索拉非尼）is a kinase inhibitor used in the treatment of liver cancer, kidney cancer（renal cell carcinoma）, and certain types of thyroid（甲状腺）cancer.

Trasylol（Aprotinin）is a trypsin（胰蛋白酶）inhibitor used to control bleeding during major surgery.

Ciprofloxacin was approved by the US Food and Drug Administration（FDA）in 1987. Ciprofloxacin is the most widely used of the second-generation quinolone（喹诺酮）antibiotics that came into clinical use in the late 1980s and early 1990s.

Rennie antacid tablets, one of the biggest selling branded over-the-counter medications sold in Great Britain, with sales of £29.8 million.

Besides, in agriculture, Bayer produces various fungicides(杀真菌剂), herbicides(除草剂), insecticides, and some crop varieties.

Fungicides are primarily marketed for cereal crops, fresh produce, fungal(真菌的) with bacteria-based pesticides, and control of mildew and rust diseases.

Herbicides are marketed primarily for field crops and orchards.

Insecticides are marketed according to specific crop and insect pest type.

(*If you want to find more information about this corporation, please log on https://en.wikipedia.org/wiki/BASF*)

1. *Read the passage quickly by using the skills of skimming and scanning. And choose the best answer to the following questions.*

   1) Bayer's primary areas of business include the following except _____.

      A. human and veterinary pharmaceuticals

      B. consumer healthcare products

      C. agricultural chemicals and biotechnology products

      D. low value polymers

   2) What is Bayer's first and best known product?

      A. Heroin.   B. Aspirin.

      C. Antibiotics.   D. Fenbid.

   3) Which is the sharing letter?

      A. B.   B. A.

      C. Y.   D. E.

   4) Who received the Nobel Prize in Medicine for this work in 1939?

      A. Fritz Mietzsch.   B. Gerhard Domagk.

      C. Joseph Klarer.   D. IG Farben.

   5) In _____, Bayer HealthCare AG (de) acquired the over-the-counter (OTC) Pharmaceutical Division of Roche Pharmaceuticals.

      A. 2004   B. 2006

      C. 2008   D. 2010

   6) Bayer Crop Science does not have products in _____.

      A. crop protection   B. nonagricultural pest control

      C. seeds and plant biotechnology   D. genetically modified food

   7) Which is not one of Bayer HealthCare's four subdivisions?

      A. Bayer Schering Pharma.   B. Bayer Consumer Care.

      C. Bayer Plant Health.   D. Bayer Medical Care.

   8) In 2014, top-selling products included the following except _____.

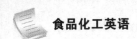

    A. Kogenate   B. Xarelto

    C. Betaseron   D. Tasmin

9) In agriculture, Bayer does not produce _____.

    A. disinfectants   B. herbicides

    C. insecticides   D. some crop varieties

10) Fungicides are primarily marketed for _____.

    A. field crops   B. orchards

    C. control of mildew and rust diseases   D. insects

2. *In this part, the students are required to make an oral presentation on either of the following topics.*

1) The secrets of Bayer's success.

2) The lessons from BASF's development history.

习题答案

Unit Ten　Food and Environment

I. Pre-class Activity

Directions: *Please read the general introduction about Adelle Davis and tell something more about the great scientist to your classmates.*

Adelle Davis

Adelle Davis was an American author and nutritionist, considered "the most famous nutritionist in the early to mid-20th century". She was as an advocate for improved health through better nutrition. She wrote an early textbook on nutrition in 1942, followed by four best-selling books for consumers which praised the value of natural foods and criticized the diet of the average American. Her books sold over 10 million copies and helped shape America's eating habits.

Despite her popularity, she was heavily criticized by her peers for many recommendations she made that were not supported by the scientific literature, many of which were considered dangerous.

Adelle Davis was born on February 25, 1904, on a small-town farm near Lizton, Indiana. She was the youngest of five daughters of Charles Eugene Davis and Harriette (McBroom) Davis.

To help her spread nutrition information to the public, she took a writing course and began writing pamphlets and books. She continued seeing patients referred to her by physicians, and by the end of her career she had helped approximately 20 000 referred patients. She had practiced professional nutritional counseling for 35 years before she gave up and devoted her time to her family.

Davis wrote her consumer books over a 40-year career, revising some in the 1970s. She saw herself as an "interpreter", not merely a researcher. "I think of myself as a newspaper reporter, who goes out to libraries and gathers information from hundreds of journals, which most people can't understand, and I write it so that people can understand." She reviews scientific literature

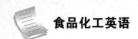

in the biochemical libraries at UCLA, for instance. Her references for *Let's Get Well* totaled almost 2 500, many from cases during her nutrition practice, and she was upset when the publisher of *Let's Have Healthy Children* eliminated the 2 000 references from the 1972 revision, says author Daniel Yergin.

II. Specialized Terms

Directions: *Please memorize the following specialized terms before the class so that you will be able to better cope with the coming tasks.*

academia n.学术界

accountability n.责任

adverse adj.不利的

affluent adj.富足的

aggravate v.恶化

alter v.改变

ambient adj.周围的

approximately adv.大约

aptitude n.资质

aquarium n.水族馆

assessment n.评估

automatic adj.自动的

be affiliated with 隶属于

bio-based 基于生物技术的

boost v.促进

booth n.展位

bulge v.鼓起

bulky adj.庞大的

byproduct n.副产品

capitalization n.资本化

carcinogenic adj.致癌的

chronically adv.慢性地

clog v.阻塞

coated adj.裹有……的

collaboration n.合作

conservation n.保护

consistent adj.一致的

contaminate v.污染

contaminated adj.污染的

contamination n.污染

coordinate v.协调

correspond v.符合

crude adj.天然的

cutting-edge adj.尖端的

dedicate v.致力于

deem v.认为

defect n.缺陷

digest v.消化

discipline n.学科

drain v.排出

droplet n.小水珠

dye n.染料

dynamic adj.充满活力的

ecosystem n.生态系统

effect v.产生

efficiency n.效率

embrocate v.涂擦

emission n.排放

entangle v.卷入

epitaxially adv.外延地

escalate v.升级,不断恶化

etching n.刻蚀

evolve v.逐步形成;发展

excel v.擅长

exfoliation n.剥落

expertise n.专门知识

expo n.博览会

extruded adj.挤压的

flatten v.变平

frigorific adj.引起寒冷的

frustration n.沮丧

fundamental adj.基本的

gamut n.全范围

generic adj.通用的

harness v.利用

highlight v.突出

hybridize v.杂交

hygroscopic adj.吸湿的

impending adj.紧迫的

imperative adj.重要的,必须处理的

impregnate v.使充满

improvised adj.临时的

incarnation n.体现,化身

inextricable adj.无法分开的

inhalational adj.吸入式的

inhale v.吸入

inherently adv.本质地

initial adj.最初的

innocuous adj.无害的

intention n.意向

invaluable adj.无价的

know-how n.专门技能

lucrative adj.赚大钱的;获利多的

margin n.空间

maximise v.使……最大化

modus operandi 工作方法

moisturize v.使……湿润

The Nature Conservancy 大自然保护协会

notification n.通告

nullify v.使无效

operational adj.运作的

oxidizing adj.氧化的

panelist n.专家

paramount adj.最高的

pathogen n.病原体

perpendicular adj.垂直的

perspiration n.汗

phase...out 逐步淘汰,逐步放弃

phenomenal adj.非凡的

pneumonia n.肺炎

III. Watching and Listening

Task One　Agriculture in America

New Words

Kansas n.美国堪萨斯州

comb n.梳子;蜂巢 v.梳理;彻底搜查

larvae n.(昆虫的)幼虫(larva 的复数)

pest n.害虫,害兽

aerial adj.航空的,空中的

deployment n.部署,调度

tunnel v.开凿隧道,挖地道

chemical n.化学制品

视频链接及文本

honeybee n.蜜蜂

suck up 吸收

nectar n.花蜜

sweeper n.打扫者;清扫器

delicate adj.微妙的;纤弱的

pollinate v.给……传授花粉

hive n.蜂箱,蜂房

arsenal n.兵工厂,军火库

Exercises

1. *Watch the video for the first time and choose the best answers to the following questions.*

1）Greg Stone's multimillion-dollar business is a _____ .

    A. aquaculture   B. polyculture

    C. monoculture   D. intercropping

2）California is the country's biggest producer of _____ .

    A. tuna   B. almond

    C. pumpkin   D. honey

3）According to David, most of bee truckers that are in pollination business are losing _____ of our bees every year.

    A. 60 to 70 percent   B. 16 to 30 percent

    C. 40 to 50 percent   D. 30 to 70 percent

4）By the end of their journey, David's bees have traveled _____ miles and pollinated an estimated _____ acres of America's produce.

    A. 7000;1500   B. 8000;5500

    C. 6500;1500   D. 8500;5000

5）If Colony Collapse Disorder keeps on happening and all the bees get wiped out, the following phenomenon will happen except that _____ .

    A. the farmers are in trouble

    B. consumers are not going to have pumpkins, apples or cherries

    C. America's food machine will feel the pressure

    D. there will be many other methods of pollination

2. *Watch the video again and decide whether the following statements are true or false.*

1）According to Robert, it's enough to use only 2 or 3 kinds of chemicals for plants in case that the yields will suffer. (　　)

2）The insects had their own secret weapon—evolution, to fight against chemicals. (　　)

3）Even high-tech farmers haven't come up with a better method for pollinating their produce without the honeybees. (　　)

4）David has to give them a kind of bee detox, because the California almond trees have been treated with tons of pesticides. (　　)

5）The scientific jury has revealed what's causing Colony Collapse Disorder for the honeybees. (　　)

3. *Watch the video for the third time and fill in the following blanks.*

    The pesticide industry has had to develop this wide ranging chemical arsenal because the insects had their own secret weapon—1）_____ . Sooner or later, the pests evolve 2）_____ to every new product. Of course, no one's been happy about this, certainly not the 3）_____ .

All right. Take a look at what we have right here. It's this beautiful 4) _____ inside this pumpkin plant. It's rolling around inside, sucking up the 5) _____. But while it's doing that, it's getting all this 6) _____ all over its legs and the rest of its body. It's a beautiful site and it shows, in action, this 7) _____ relationship between the honeybee and the 8) _____ itself. That's absolutely critical to making sure that this turns into that. Even high-tech farmers haven't 9) _____ a better method for pollinating their 10) _____. But there aren't enough wild honeybees to do the job.

4. *Share your opinions with your partners on the following topics for discussion.*

1) The agricultural difference between China and America.

2) Your suggestion on having a healthy diet and keeping a sustainable environment at the same time.

Task Two   Air Pollution

New Words

visible adj.明显的;看得见的

smog n.烟雾

prevail v.盛行,流行

gritty adj.坚韧不拔的

smut n.[植保]黑穗病;煤尘;污迹

Clean Air Act 清洁空气法

heavy oil 重油

maritime adj.海的,海事的,海运的;近海区的

negligible adj.微不足道的

视频链接及文本

Exercises

1. *Watch the video for the first time and choose the best answers to the following questions.*

1) The London smog of 1952 caused the death of _____ people.

    A. 3000   B. 3500

    C. 4000   D. 4500

2) Particles are derived from _____ to a very large extent.

    A. planes   B. vehicles

    C. factories   D. ships

3) Which of the following is not the source of pollution?

    A. Cars and lorries.   B. Factories.

    C. Power stations.   D. Bus stations.

4) The Loyd's Register's data about air pollution was found to be _____.

    A. old   B. new

    C. unique   D. inspiring

5) Loyd is asked to comment on _____ papers.

    A. one   B. two

    C. three   D. four

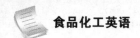

2. *Watch the video again and decide whether the following statements are true or false.*

   1）The London smog of 1952 led to the 1954 *Clean Air Act.*（ ）

   2）We blame cars and trucks for all of the pollution.（ ）

   3）Today's emissions are more subtle and invisible.（ ）

   4）We've come to expect the air in the sea to be very healthy.（ ）

   5）Two papers submitted to the International Maritime Organization lead to two opposite conclusions.（ ）

3. *Watch the video for the third time and fill in the following blanks.*

   Air pollution was once all too 1）＿＿＿＿ in the cities of Britain. Dense 2）＿＿＿＿ prevailed places like London, causing severe health problems in the growing 3）＿＿＿＿ world. The London smog of 1952 caused the death of 4）＿＿＿＿ people and led to the 5）＿＿＿＿ *Clean Air Act.* In the London of today in cities around the world, such 6）＿＿＿＿ are considered unacceptably 7）＿＿＿＿. But we know that the 8）＿＿＿＿ has not gone away. It's just become 9）＿＿＿＿ visible.

4. *Share your opinions with your partners on the following topics for discussion.*

   1）What do you think about air pollution?

   2）Do you have any ideas about specific measures of protecting the environment?

IV. Talking

Task One   Classical Sentences

Directions： *In this section, some popular sentences are supplied for you to read and to memorize. Then, you are required to simulate and produce your own sentences with reference to the structure.*

General Sentences

1. What channel did you watch on television last night?
   昨天晚上你看的哪个频道的电视?

2. I don't get a good picture on my TV set. There's probably something wrong.
   我的电视机上画面不清楚,可能出毛病了。

3. You get good reception on your radio.
   你的收音机接收效果很好。

4. Please turn the radio up. It's too low.
   请把收音机开大点声,声音太小了。

5. What's on following the news and weather?
   新闻和天气预报后是什么节目?

6. Do you have a TV guide?
   你有电视节目指南吗?

7. You ought to have Bill look at your TV. Maybe he could fix it.
   你应该让比尔检查下你的电视,他或许能修好。

8. We met one of the engineers over at the television station.
   我们在电视台遇见了一个在那里工作的工程师。

9. Where can I plug in the TV? Is this outlet all right?
   电视插头该插在哪里? 这个插头可以用吗?

10. I couldn't hear the program because there was too much static.
    因为干扰太大,我听不清节目了。

11. Your car radio works very well. What kind is it?
    你的车载收音机性能很好,它是什么类型的?

12. Next time I buy a TV set, I'm going to buy a portable model.
    下一次我买电视机时,我打算买一部手提式的。

13. I wonder if this is a local broadcast.
    我想知道这是不是本地广播。

14. You'd get better TV reception if you had an outside antenna.
    如果你有室外天线的话,你电视机的接收效果将会很好。

15. Most amateur radio operators build their own equipment.
    大多数业余收音机爱好者都自己组装收音设备。

16. Station Voice of WIT is off the air now. They signed off two hours ago.
    武工之声电台已经停止广播了,它们在两小时之前就结束了。

17. Who is the author of this novel?
    这个小说的作者是谁?

18. I've never read a more stirring story.
    我从没有读过这么感人的故事。

19. Who would you name as the greatest poet of our time?
    你认为谁是我们这个时代最伟大的诗人?

20. This poetry is realistic. I don't care for it very much.
    这篇诗集是现实主义的,我不太喜欢。

21. Many great writers were not appreciated fully while they were alive.
    许多大作家在世时并没有得到人们的充分赏识。

22. This is a poem about frontier life in the United States.
    这是一首描述美国边境生活的诗。

23. This writer uses vivid descriptions in his writings.
    这位作者作品中的描述非常生动。

24. How much do you know about the works of Henry Wadsworth Longfellow?
    关于亨利·沃兹沃思·朗费罗的著作,你了解多少?

25. As the saying goes, "Where there is a will, there is a way".
    俗话说,有志者事竟成。

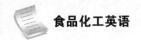

26. It is well-known to all that all roads lead to Rome.
众所周知,条条大路通罗马。

27. Whatever is worth doing is worth doing well.
任何值得做的,就要把它做好。

28. The hardest thing to learn is to be a good loser.
最难学的是做一个输得起的人。

29. Happiness is a way station between too much and too little.
幸福,是位于太多和太少之间的一个小站。

30. The hard part isn't making the decision. It's living with it.
做出决定并不困难,困难的是接受决定。

31. You may be out of my sight, but never out of my mind.
你也许已走出我的视线,但从未走出我的思念。

32. Love is not a maybe thing. You know when you love someone.
爱不是什么可能、大概、也许,一旦爱上了,自己是十分清楚的。

33. In the end, it's not the years in your life that count. It's the life in your years.
到头来,你活了多少岁不算什么,重要的是,你是如何度过这些岁月的。

34. When the whole world is about to rain, let's make it clear in our heart together.
当全世界约好一起下雨,让我们约好一起在心里放晴。

35. It's better to be alone than to be with someone you're not happy to be with.
宁愿一个人呆着,也不要跟不合拍的人呆一块。

36. One needs three things to be truly happy living in the world: something to do, someone to love, and something to hope for.
要得到真正的快乐,我们只需拥有三样东西:想做的事、值得爱的人、美丽的梦。

37. No matter how badly your heart has been broken, the world doesn't stop for your grief. The sun comes right back up the next day.
不管你有多痛苦,这个世界都不会为你停止转动,太阳照样升起。

38. Accept what was and what is, and you'll have more positive energy to pursue what will be.
接受过去和现在的模样,才会有能量去追寻自己的未来。

39. Until you make peace with who you are, you'll never be content with what you have.
除非你能和真实的自己和平相处,否则你永远不会对已拥有的东西感到满足。

40. If you wish to succeed, you should use persistence as your good friend, experience as your reference, prudence as your brother and hope as your sentry.
如果你希望成功,当以恒心为良友,以经验为参谋,以谨慎为兄弟,以希望为哨兵。

41. Great minds have purpose, others have wishes.
杰出的人有着目标,其他人只有愿望。

42. Being single is better than being in an unfaithful relationship.
比起谈着充满欺骗的恋爱,单身反而更好。

43. If you find a path with no obstacles, it probably doesn't lead anywhere.
太容易的路,可能根本就不会带你去任何地方。

44. Getting out of bed in winter is one of life's hardest mission.
冬天起床是人生最艰难的任务之一。

45. The future is scary but you can't just run to the past because it's familiar.
未来会让人心生畏惧,但是我们却不能因为习惯了过去,就逃回过去。

46. Success is the ability to go from one failure to another with no loss of enthusiasm.
成功是,你即使跨过一个又一个失败,但也没有失去热情。

47. Not everything that is faced can be changed, but nothing can be changed until it is faced.
并不是你面对了,事情都能改变。但是,如果你不肯面对,那什么也变不了。

48. If they throw stones at you, don't throw back. Use them to build your own foundation instead.
如果别人朝你扔石头,就不要扔回去了,留着作你建高楼的基石吧!

49. If your happiness depends on what somebody else does, I guess you do have a problem.
如果你的快乐与否取决于别人做了什么,我想,你真的有点问题。

50. Today, give a stranger one of your smiles. It might be the only sunshine he sees all day.
今天,给一个陌生人送上你的微笑吧,很可能,这是他一天中见到的唯一的阳光。

Specialized Sentences

1. It is better to prevent waste than to treat or clean up waste after it has been created.
与其制造垃圾后清理,不如少制造。

2. Chemical products should be designed to realize their desired function while minimizing their toxicity.
人们设计化学产品时,要达到它们理想的功能,同时也要最大限度地减少毒性。

3. A raw material should be renewable and practical.
原材料应该实用,且可循环利用。

4. This will give a competitive advantage to those companies that promote green chemical products.
那些提倡绿色化学产品的公司会拥有竞争优势。

5. Governments have a major role in adopting policies that promote renewable energy.
在采取提倡可再生能源的政策方面,政府发挥着重要的作用。

6. Manufacturers and retailers have a responsibility to demand safe chemicals from their suppliers.
制造商和零售商有责任要求供应商提供安全的化工品。

7. All sides seek to reduce and prevent pollution at its source.
各界应从源头寻求减少、预防污染的措施。

8. The key is on minimizing the hazard and maximizing the effectiveness.
重点是减少危害,提高实效。

9. Attempts are being made to protect the environment and guarantee the safety of food.

人们努力保护环境,并保证食品安全。

10. Protecting the environment is increasingly seen as a powerful concept that gains the trends at present.

保护环境日益成为一个当下时兴的重要理念。

11. It aims to avoid problems before they happen.

它旨在未雨绸缪。

12. New products minimize the use and generation of hazardous substances.

新的产品减少了有毒物质的使用和生产。

13. A variety of food also causes potential dangers.

各种各样的食物也会引起潜在的危险。

14. Global warming and pollution may result in more food safety problems.

全球变暖和污染会导致更多的食品安全问题。

15. Examples of mad cow and foot and mouth diseases have reminded the world of the importance of information sharing and crisis warning system.

疯牛病和手足口病的例子,提醒全世界人民信息共享和危机预警系统的重要性。

16. Intensified international cooperation is a solution to the problem.

日益紧密的国际合作提供了一个问题解决方式。

17. The effects of air pollution are influenced by the type and quantity of pollutants.

污染物的类型和数量决定空气污染的后果。

18. Air pollution is recognized as a source of discomfort for centuries as smoke, dust and obnoxious odors.

几百年来,人们一直认为空气污染、烟尘和有毒气体是引发不适的原因。

19. The illnesses caused by air pollution are characterized by cough and sore throat, irritation of the eyes, nose, throat and respiratory tract, plus stress on the heart.

由空气污染引发的疾病的特点有咳嗽,嗓子疼,眼睛、鼻子、喉咙和呼吸系统的瘙痒以及心脏的压力。

20. Protecting the public health is of great necessity.

保护公众健康很重要。

21. The type of air pollutants are related to the original material used for processing.

空气污染物的类型与加工的原材料息息相关。

22. The source of nitrogen is principally the air used in combustion.

氮的来源主要是燃烧过程中使用的空气。

23. Progress has been made in the control of sulfur dioxide and particulates.

二氧化硫和微粒的控制,取得长足的进步。

24. Greater emphasis is shifting to the control of sulfates and nitrates.

人们变得更重视对硫酸盐和硝酸盐的控制。

25. The rapid growth in industrial production during the last decades has enhanced the industrial air problem.

过去几十年,工业生产的快速发展加剧了工业空气问题。

26. A distinction should be made between air quality standards.

空气质量标准应有所区分。

27. Industrial air pollution covers a wide range of problems.

工业空气污染包含很多问题。

28. All the methods are equally valid for industrial air pollution control.

所有办法对于工业空气污染的控制同样有效。

29. Living environment is a big issue nowadays.

生存环境是当今的一个大问题。

30. Forests are an important habitat for thousands of species of animal and plant life.

森林是成千上万种动物和植物的重要栖息地。

31. Environmental problems can undermine the quality of people's life.

环境问题危害人类的生活质量。

32. It will need much more money and energy to repair the environment if it is polluted.

环境一旦污染,就要花费更多金钱和精力来修复。

33. We should seek a balance between environmental protection and economic growth.

我们应该平衡环境保护和经济增长之间的关系。

34. The application of green chemistry is becoming more widespread.

绿色化学的应用越发普遍。

35. We should put more emphasis on environmental protection than economic development.

我们应该更强调环境保护,而不是经济发展。

36. We ought to look at the environmental issue in the long term.

我们应该从长远的角度看待环境问题。

37. Environmental cost should be taken into consideration when developing economy.

发展经济时,要考虑环境成本。

38. It is increasingly acknowledged that protecting the environment is the most important goal.

人们普遍认为,保护环境是最重要的目标。

39. Bioengineering is seen as a promising technique for achieving the green chemistry goal.

生物工程是实现绿色化学目标的有前途的技术手段。

40. Both the environmental and human health impacts of the products themselves must be considered to ensure their long term economic viability.

必需考虑产品本身对环境和人类健康的影响,以保证长足的经济可行性。

41. Laws and regulations help create ways for dealing with pollution in an innovative way.

法律法规有助于找到解决污染问题的新方法。

42. Some tasty food can have adverse health consequences.

一些美味的食物会给健康带来负面后果。

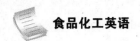

43. How should we alert the public to the hazard of air pollution?

我们如何提醒公众空气污染的危害?

44. This is why it is critical to spend time popularizing the importance of protecting the environment.

这是为什么要花时间普及保护环境的重要性。

45. The destructions may manifest themselves in future years.

这种毁坏在今后会显现出来。

46. The law requires companies to make publicly available information about their process.

法律要求公司公开生产过程的有效信息。

47. Food poisoning warns people of potential hazards resulting from unknown food.

食物中毒警告人们,小心未知食物的可能危害。

48. Appropriate measures are needed to enhance workers' awareness against unlawful process.

人们要采取适当措施,以提高工人抵制非法生产过程的意识。

49. Failure to provide proper protection results in greater dangers.

不能提供合适的保护,会导致更严重的危险。

50. More attention should be paid to sustainable development while promoting economic surge.

在发展经济的同时,人们更应关注可持续发展。

Task Two　Sample Dialogue

Directions：*In this section, you are going to read several times the following sample dialogue about the relevant topic. Please pay special attention to five C's (culture, context, coherence, cohesion and critique) in the dialogue and get ready for a smooth communication in the coming task.*

Talking about air pollution

Jones：I find that protecting the environment becomes a critical issue nowadays.

Mike：I think so. There are more and more vehicles on the road. The waste gas leads to the serious pollution in the atmosphere.

Jones：Besides, the rapid industrial growth like factories contributes to the air pollution.

Mike：It seems that governments pay more attention to the economic development than environmental protection.

Jones：It's true. Everyone is looking at the issue from their own perspective in the short term. Countries and companies just want to take advantage of nature to make money.

Mike：Several disasters have alerted people to the importance of protecting the environment.

Jones：We should remember that keeping a clean habitat is very important for thousands of species of animal and plant life.

Mike：What measures do you suggest?

Jones: Well, it needs joint efforts from all sides. Governments and companies should take the lead in stopping pollution at its source. Personally, I would like to join an organization committed to protecting the environment. In this way, I could get involved in projects to improve the environment.

Mike: We could apply what we have learned to the work of protecting the environment after graduation.

Task Three Simulation and Reproduction

Directions: *The class will be divided into two major groups, each of which will be assigned a topic. In the process of discussion, please observe the principles of cooperation, politeness and choice of words. One of the groups will be chosen to demonstrate the discussion to the class.*

1) Water pollution

2) Land pollution

Task Four Discussion and Debate

Directions: *The class will be divided into two groups. Please choose your stand in regard to the following controversy and support your opinions with scientific evidences. Please refer to the specialized terms and classical sentences in the previous parts of this unit.*

There is no doubt that economic development is very critical to the improvement of people's living standard. It is believed by many that boosting the economy should be more important than protecting the environment. However, others believe that the environment should be given priority over others. Which party do you agree with? Why?

V. After-class Exercises

1. *Match the English words in Column A with the Chinese meaning in Column B.*

A	B
1) pollutant	a. 医师
2) byproduct	b. 蜂巢
3) kidney	c. 可行性
4) ecosystem	d. 混养
5) polyculture	e. 微不足道的
6) viability	f. 昏迷
7) honeycomb	g. 副产品
8) coma	h. 肾脏
9) medic	i. 生态系统
10) negligible	j. 污染物

2. *Fill in the following blanks with the words or phrases in the word bank. Change the forms if it's necessary.*

venom	larvae	toxin	prioritize	assessment
paramount	mercury	inhale	famine	accountability
boost	dissolve	addictive	contamination	byproduct

1) Phosphates in themselves are not _____（毒素）.

2) The Environmental Protection Agency recognized early the problem of phosphorus pollution and has proactively addressed phosphate _____（污染）.

3) Cellulose materials _____（溶解）and become biodegradable in nature.

4) Phosphates are a common _____（添加剂）in dish soaps.

5) Chlorine can generate _____（副产品）such as hydrochloric acid.

6) Measures are taken to deal with _____（污染）.

7) Clean energy is _____（居于优先地位）in terms of the conventional fuel.

8) Protecting the environment is of _____（主要的）importance.

9) People died of _____（饥荒）and diseases.

10) _____（水银）could be fatal when it is taken into the body.

3. *Translate the following paragraph into Chinese.*

　　Air pollution occurs when harmful or excessive quantities of substances including gases, particulates, and biological molecules are introduced into the Earth's atmosphere. It may cause diseases, allergies and even death to humans; it may also cause harm to other living organisms, such as animals and food crops, and may damage the natural or built environment. Both human activities and natural processes can generate air pollution. According to the 2014 World Health Organization report, air pollution in 2012 caused the deaths of around 7 million people worldwide.

_____

_____

_____

_____

_____

_____

_____

4. *Translate the following paragraph into English.*

　　环境保护已成为当今世界各国政府和人民的共同行动和主要任务之一。我国则把环境保护宣布为我国的一项基本国策，并制定和颁布了一系列环境保护的法律、法规以保证这一基本国策的贯彻执行。随着经济和贸易的全球化,环境污染也日益呈现国际化趋势,危险废物越境转移问题就是这方面的突出表现。

_____

_____

_____

_____

_____

_____

_____

5. *Please write an essay of about 120 words on the topic* "**Protecting the Environment in Our Life**". *Some specific examples will be highly appreciated and watch out the spelling of some specialized terms you have learnt in this unit.*

_____

_____

_____

_____

_____

_____

_____

_____

_____

_____

VI. Additional Reading

A Brief Introduction on DuPont

E. I. du Pont de Nemours and Company, commonly referred to as DuPont, is an American conglomerate (联合大企业) that was founded in July 1802 as a gunpowder mill(火药作坊) by French-American chemist and industrialist Éleuthère Irénée du Pont.

In the 20th century, DuPont developed many polymers(聚合物) such as Vespel, neoprene

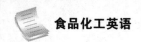

（氯丁橡胶）, nylon, Corian, Teflon, Mylar, Kapton, Kevlar, Zemdrain, M5 fiber, Nomex, Tyvek, Sorona, Corfam, and Lycra. DuPont developed freon（氟利昂）（chlorofluorocarbons, 氯氟碳化物）for the refrigerant（制冷剂）industry, and later more environmentally friendly refrigerants. It also developed synthetic pigments（合成色素）and paints including ChromaFlair（珂玛菲）.

In 2014, DuPont was the world's fourth largest chemical company based on market capitalization（市场资本化）and eighth largest based on revenue（收入）. On August 31, 2017, it merged（合并）with the Dow Chemical Company to create DowDuPont, the world's largest chemical company in terms of sales, of which DuPont is now a subsidiary（子公司）. Its stock price is a component of the Dow Jones Industrial Average（道琼斯工业平均指数）.

Core Values（核心价值观）

Our core values are the cornerstone（基石）of who we are, what we stand for and what we do. DuPont businesses help provide safe, sufficient food; ample（富余的）, sustainable（可持续的）energy; and protection for people and the environment. Even as our company grows and the Earth's population surges（快速增长）, we have never changed the commitment we share to our core values:

1. Safety and Health

We share a personal and professional commitment to protecting the safety and health of our employees, our contractors（承包商）, our customers and the people of the communities in which we operate.

2. Environmental Stewardship（管理）

We find science-enabled, sustainable solutions for our customers, always managing our businesses to protect the environment and preserve the earth's natural resources, both for today and for generations into the future.

3. Respect for People

We treat our employees and all our partners with professionalism（职业精神）, dignity and respect, fostering an environment where people can contribute, innovate（创新）and excel（卓越）.

4. Highest Ethical（伦理）Behavior

We conduct ourselves and our business affairs in accordance with the highest ethical standards, and in compliance with all applicable laws, striving always to be a respected corporate citizen worldwide.

DuPont Code of Conduct（行为准则）

The DuPont Code of Conduct provides information:

To guide employees so that their business conduct is consistent（一致的）with the company's ethical standards.

Whenever you have questions or concerns regarding an ethical situation, you are encouraged to discuss these concerns with local line management. If you do not feel comfortable discussing the concerns face-to-face, you can call the DuPont Ethics and Compliance Hotline（道德操守及

合规热线). Such situations may include：

Theft(盗窃)or fraud(欺骗)

Accounting(财会)or auditing(审计)issues

Conflicts of interest

Inappropriate business gifts

Safety, health, or environmental issues

Misuse of company assets(滥用公司资产)

Non-compliance with laws

The DuPont Ethics and Compliance Hotline is a multi-lingual, free phone number to call to report suspected violations(违反) of our Code of Conduct. The hotline is available 24 hours a day, 7 days a week. The callers can choose to remain anonymous(匿名的). If you call the hotline, a trained specialist, who is employed by an outside firm, will listen to your concerns and take notes to prepare a report that will be forwarded(递交) to the appropriate DuPont management for review, or, in cases involving auditing, accounting or internal controls issues, to the Audit Committee of the DuPont Board of Directors(董事会).

People who wish to submit their concern using the internet can do so through a website. This site will allow you to create an unique user ID and set up a password to submit a confidential (保密的) concern to the third-party provider. The concern will be handled following the same process that is in place for hotline calls.

To view these files, you will need the Adobe Acrobat Reader. Most browsers today are equipped with Acrobat, but if you cannot view these documents, a free reader is available for download.

History

1. Establishment：1802

DuPont was founded in 1802 by Éleuthère Irénée du Pont, using capital raised in France and gunpowder machinery(机械) imported from France. The company was started at the Eleutherian Mills, on the Brandywine Creek, near Wilmington, Delaware, two years after he and his family left France to escape the French Revolution and religious persecutions(迫害) against Huguenot Protestants(新教徒). The company began as a manufacturer of gunpowder, as du Pont noticed that the industry in North America was lagging behind(滞后) Europe. The company grew quickly, and by the mid-19th century had become the largest supplier of gunpowder to the United States military, supplying half the powder used by the Union Army during the American Civil War. The Eleutherian Mills site is now a museum and a national historic landmark(地标).

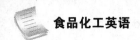

2. Expansion: 1902 to 1912

DuPont continued to expand, moving into the production of dynamite(硝化甘油炸药)and smokeless powder. In 1902, DuPont's president, Eugene du Pont, died, and the surviving partners sold the company to three great-grandsons of the original founder. Charles Lee Reese was appointed as director and the company began centralizing(集中)their research departments. The company subsequently purchased several smaller chemical companies, and in 1912 these actions gave rise to government scrutiny(仔细检查)under the Sherman Antitrust Act(谢尔曼反托拉斯法). The courts declared that the company's dominance(主宰)of the explosives business constituted a monopoly(垄断)and ordered divestment(撤资). The court ruling(裁决)resulted in the creation of the Hercules Powder Company(later Hercules Inc. and now part of Ashland Inc.)and the Atlas Powder Company［purchased by Imperial Chemical Industries(ICI)and now part of AkzoNobel］. At the time of divestment, DuPont retained(保留)the single base nitrocellulose(硝化纤维素)powders, while Hercules held the double base powders combining nitrocellulose and nitroglycerine(硝化甘油). DuPont subsequently developed the Improved Military Rifle(IMR,步枪)line of smokeless powders.

In 1910, DuPont published a brochure(手册)entitled "Farming with Dynamite". The pamphlet(小册子)was instructional, outlining the benefits to using their dynamite products on stumps(树桩)and various other obstacles that would be easier to remove with dynamite as opposed to other more conventional and inefficient means.

DuPont also established two of the first industrial laboratories in the United States, where they began the work on cellulose(纤维素)chemistry, lacquers(漆)and other non-explosive products. DuPont Central Research was established at the DuPont Experimental Station, across the Brandywine Creek from the original powder mills.

3. Automotive(汽车的) investments: 1914

In 1914, Pierre S. du Pont invested in the fledgling(新兴的)automobile industry, buying stock in General Motors(GM,通用汽车公司). The following year he was invited to be on GM's board of directors and would eventually be appointed the company's chairman. The DuPont Company would assist the struggling automobile company further with a $25 million purchase of GM stock. In 1920, Pierre S. du Pont was elected president of General Motors. Under du Pont's guidance, GM became the number one automobile company in the world. However, in 1957, because of DuPont's influence within GM, further action under the Clayton Antitrust Act(克雷顿反托拉斯法)forced DuPont to divest its shares of General Motors.

4. Major breakthroughs: 1920s to 1930s

In the 1920s, DuPont continued its emphasis on materials science, hiring Wallace Carothers to work on polymers in 1928. Carothers invented neoprene, a synthetic rubber; the first polyester superpolymer(聚酯超聚合物); and, in 1935, nylon. The invention of Teflon followed a few years later. DuPont introduced phenothiazine(吩噻嗪)as an insecticide(杀虫剂)in 1935.

5. Second World War: 1941 to 1945

DuPont ranked 15th among United States corporations in the value of wartime production contracts. As the inventor and manufacturer of nylon, DuPont helped produce the raw materials for parachutes(降落伞), powder bags, and tires.

DuPont also played a major role in the Manhattan Project in 1943, designing, building and operating the Hanford plutonium(钚)producing plant in Hanford, Washington. In 1950, DuPont also agreed to build the Savannah River Plant in South Carolina as part of the effort to create a hydrogen bomb.

6. Space Age developments: 1950 to 1970

After the war, DuPont continued its emphasis on new materials, developing Mylar, Dacron, Orlon, and Lycra in the 1950s, and Tyvek, Nomex, Qiana, Corfam, and Corian in the 1960s. DuPont materials were critical to the success of the Apollo Project of the United States space program.

DuPont has been the key company behind the development of modern body armor(盔甲). In the Second World War, DuPont's ballistic(弹道学的) nylon was used by Britain's Royal Air Force to make flak jackets（防弹背心）. With the development of Kevlar in the 1960s, DuPont began tests to see if it could resist a lead bullet. This research would ultimately lead to the bullet-resistant vests that are the mainstay（支柱）of police and military units in the industrialized world.

A marker outside DuPont's Belle Plant in Dupont City, West Virginia, where ammonia (氨) was first synthesized for commercial use

7. Conoco holdings: 1981 to 1999

DuPont's Orlon plant in Camden, South Carolina, 1930–1945

In 1981, DuPont acquired Conoco Inc., a major American oil and gas producing company that gave it a secure source of petroleum feedstocks(原料) needed for the manufacturing of many of its fiber and plastics products. The acquisition, which made DuPont one of the top ten US-based petroleum and natural gas producers and refiners(炼油公司), came about after a bidding war with the giant distillery（蒸馏酿酒厂）Seagram Company Ltd., which would become DuPont's largest single shareholder with

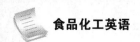
four seats on the board of directors. On April 6,1995, after being approached by Seagram Chief Executive Officer Edgar Bronfman Jr., DuPont announced a deal in which the company would buy back all the shares owned by Seagram.

In 1999, DuPont sold all of its shares of Conoco, which merged with Phillips Petroleum Company, and acquired the Pioneer Hi-Bred agricultural seed company.

8. Activities, 2000–present

DuPont describes itself as a global science company that employs more than 60 000 people worldwide and has a diverse(多样的) array(分布) of product offerings. The company ranks 86th in the Fortune 500 on the strength of nearly $36 billion in revenues and $4.848 billion in profits in 2013. In April 2014, *Forbes* ranked DuPont 171st on its Global 2000, the listing of the world's top public companies.

DuPont businesses are organized into the following five categories, known as marketing "platforms": Electronic and Communication Technologies, Performance Materials, Coatings(涂层) and Color Technologies, Safety and Protection, and Agriculture and Nutrition.

The agriculture division, DuPont Pioneer makes and sells hybrid(杂交) seed and genetically modified(转基因) seed, some of which goes on to become genetically modified food. Genes engineered into their products include LibertyLink, which provides resistance to Bayer's Ignite Herbicide/Liberty herbicides; the Herculex I Insect Protection gene which provides protection against various insects; the Herculex RW insect protection trait which provides protection against other insects; the YieldGard Corn Borer gene, which provides resistance to another set of insects; and the Roundup Ready Corn 2 trait that provides crop resistance against glyphosate herbicides(草甘膦除草剂). In 2010, DuPont Pioneer received approval to start marketing Plenish soybeans(大豆油), which contain "the highest oleic acid(油酸) content of any commercial soybean product, at more than 75 percent. Plenish provides a product with no trans fat, 20 percent less saturated fat(饱和脂肪) than regular soybean oil, and more stabile oil with greater flexibility in food and industrial applications". Plenish is genetically engineered to "block the formation of enzymes(酶) that continue the cascade(瀑布) downstream from oleic acid (that produces saturated fats), resulting in an accumulation of the desirable monounsaturated acid(单不饱和酸)".

In October 2001, the company sold its pharmaceutical(制药的) business to Bristol Myers Squibb for $7.798 billion.

In 2002, the company sold the Clysar(R) business to Bemis Company for $143 million.

In 2004, the company sold its textiles(纺织品) business, which included some of its best-known brands such as Lycra(Spandex), Dacron polyester(聚酯纤维), Orlon acrylic(丙烯酸), Antron nylon and Thermolite, to Koch Industries.

In 2011, DuPont was the largest producer of titanium dioxide(二氧化钛) in the world, primarily provided as a white pigment(色素) used in the paper industry.

DuPont has 150 research and development facilities located in China, Brazil, India, Germany, and Switzerland with an average investment of $2 billion annually(每年地) in a diverse range of technologies for many markets including agriculture, genetic traits(遗传性状), biofuels(生物燃料), automotive, construction, electronics, chemicals, and industrial materials. DuPont employs more than 10 000 scientists and engineers around the world.

On January 9, 2011, DuPont announced that it had reached an agreement to buy Danish company Danisco for US $6.3 billion. On May 16, 2011, DuPont announced that its tender offer for Danisco had been successful and that it would proceed to redeem(赎回) the remaining shares and delist(退市) the company.

On May 1, 2012, DuPont announced that it had acquired from Bunge full ownership of the Solae joint venture, a soy-based ingredients(成分) company. DuPont previously owned 72 percent of the joint venture while Bunge owned the remaining 28 percent.

In February 2013, DuPont Performance Coatings was sold to the Carlyle Group and rebranded (重新获得品牌) as Axalta Coating Systems.

In October 2015, DuPont sold the Neoprene chloroprene rubber business to Denka Performance Elastomers, a joint venture of Denka and Mitsui.

9. Chemours

In October 2013, DuPont announced that it was planning to spin off(母公司收回子公司全部股本使之脱离的做法) its Performance Chemicals business into a new publicly traded company in mid-2015. The company filed its initial Form 10 with the SEC in December 2014 and announced that the new company would be called The Chemours Company. The spin-off to DuPont shareholders was completed on July 1, 2015, and Chemours stock began trading on the New York Stock Exchange on the same date.

DuPont will focus on production of GMO seeds, materials for solar panels(太阳能板), and alternatives(替代品) to fossil fuels(化石燃料). Chemours becomes responsible for the cleanup of 171 former DuPont sites, which DuPont says will cost between $295 million and $945 million.

10. Merger with Dow

On December 11, 2015, DuPont announced that it would merge with the Dow Chemical Company, in an all-stock deal(全股交易). The combined company, which will be known as DowDuPont, will have an estimated value of $130 billion, be equally held by the shareholders of both companies, and maintain their headquarters in Delaware and Michigan, respectively. Within two years of the merger's closure(关闭), expected in the first quarter of 2017 and subject to regulatory approval, DowDuPont will be split into three separate public companies, focusing on the agricultural chemicals, materials science, and specialty product industries. Commentators(评论员) have questioned the economic viability(可行性) of this plan because, of the three companies, only the specialty products industry has prospects for high growth. The outlook on the profitability(获利) of the other two proposed companies has been questioned due to reduced crop prices and lower margins(边界) on plastics such as polyethylene(聚乙烯).

They have also noted that the deal is likely to face antitrust scrutiny(反托拉斯审查)in several countries. This eventually became the case, with two delays taking place due to regulatory approvals. The merger closed on August 31,2017.

(*If you want to find more information about this corporation, please log on https://en.wikipedia. org/wiki/DuPont*)

1. *Read the passage quickly by using the skills of skimming and scanning. And choose the best answer to the following questions.*

1) When DuPont was founded in 1802, its main product was _____.

    A. refrigerant   B. pigment

    C. gunpowder   D. pesticide

2) Which of the following is not included in the core values of DuPont?

    A. Safety and Health.   B. Environmental Management.

    C. Respect for People.   D. World Peace.

3) The staff member can call the hotline if all the following accidents except _____ occur in DuPont.

    A. accounting or auditing issues

    B. disrespect for the staff members

    C. safety, health, or environmental issues

    D. misuse of company assets

4) In 1914, Pierre S. du Pont invested in the fledgling automobile industry, buying stock in _____.

    A. General Motors(GM)   B. Ford Motors

    C. Nissan Motors   D. Volkswagon Motors

5) During World War II, as the inventor and manufacturer of nylon, DuPont helped produce the raw materials for all the following except _____.

    A. parachutes   B. powder bags

    C. tires   D. flak jackets

6) The acquisition of _____ made DuPont one of the top ten US-based petroleum and natural gas producers and refiners.

    A. Conoco Inc.   B. Seagram Company Ltd.

    C. Phillips Petroleum Company   D. Bristol Myers Squibb

7) The company ranks _____ in the Fortune 500 on the strength of nearly $36 billion in revenues.

    A. 4848th   B. 171st

    C. 2000th   D. 86th

8) Which of the following is not one of the marketing platforms in DuPont?

    A. Electronic and Communication Technologies.

B. Agriculture and Nutrition.

  C. Computer Science.

  D. Coatings and Color Technologies.

9) In 2002, the company sold _____ to Bemis Company for $143 million.

  A. its pharmaceutical business

  B. the Clysar(R) business

  C. its textile business

  D. the rubber business

10) After the merging with _____, DuPont got a new name—DowDuPont.

  A. Dow Chemical Company

  B. Chemours Company

  C. Carlyle Group

  D. Denka Performance Elastomers

2. *In this part, the students are required to make an oral presentation on either of the following topics. When necessary, the students can log onto the Internet to search for relevant information.*

1) The founder of DuPont.

2) The art of "give and take" — DuPont's merging and being merged.

习题答案